NATURAL
INSECT
CONTROL

THE ECOLOGICAL

GARDENER'S GUIDE

TO FOILING PESTS

FOR THE
ADVANCE
MENT OF
BOTANY
AND THE
SERVICE OF
THE CITY

BROOKLYN
BOTANIC
GARDEN
PUBLICATIONS
· MCMXCIV ·

Janet Marinelli
EDITOR

Bekka Lindstrom
ART DIRECTOR

Barbara B. Pesch
DIRECTOR OF PUBLICATIONS

Stephen K-M. Tim
VICE PRESIDENT, SCIENCE & PUBLICATIONS

Judith D. Zuk
PRESIDENT

Elizabeth Scholtz
DIRECTOR EMERITUS

NATURAL INSECT CONTROL

THE

ECOLOGICAL

GARDENER'S

GUIDE TO

FOILING

PESTS

Warren Schultz ⁄ Guest Editor

Steve Buchanan ⁄ Illustrator

Handbook #139

BBG gardening guides are published quarterly at 1000 Washington Ave., Brooklyn, NY 11225

Subscription included in Brooklyn Botanic Garden membership dues ($25.00 per year)

ISSN 0362-5850 ISBN # 0-945352-83-2

Table of Contents

CONTRIBUTORS

LINDA AMEROSO, an entomologist with Cornell Cooperative Extension, wrote the section on the hemlock woolly adelgid. She also served as technical consultant for this book.

WHITNEY CRANSHAW wrote the chapter on natural pesticides and contributed the sections on apple maggots, coddling moths, peach tree borers, pear psyllas and plum curculios. He is an associate professor and Extension entomologist at Colorado State University in Fort Collins, Colorado, and author of *Pests of the West*.

PATRICIA KITE contributed the sections on cabbage loopers, cabbageworms, leaf miners, squash bugs and squash vine borers. She is the author of *Controlling Lawn and Garden Insects* and contributing author of *Home Gardener's Problem Solver*, both published by Ortho books.

CHARLIE NARDOZZI contributed the sections on cabbage maggots, carrot rust flies, nematodes, northern corn root worms and onion maggots. He is a gardening radio show host and horticulturist at the National Gardening Association in Burlington, Vermont.

RITA PELCZAR contributed the sections on Japanese beetles, June beetles, lace bugs, rose chafers and weevils. She is a freelance writer who has contributed to *Horticulture, American Horticulturist, National Gardening* and *Fine Gardening*.

PAM PIERCE contributed the sections on earwigs, nematodes, snails and slugs, sowbugs and pillbugs and wireworms. She has written several books. She also teaches horticulture at City College of San Francisco and is a founding board member of the San Francisco League of Urban Gardeners (SLUG).

JOANNA PONCAVAGE wrote the chapter on beneficial insects and contributed the sections on Colorado potato beetles, cucumber beetles, flea beetles, Mexican bean beetles and potato and beet leafhoppers. She is a senior editor of *Organic Gardening*.

JACK RUTTLE wrote the chapter on physical barriers and cultural controls. He has a large garden of food plants and flowers near Philadelphia. He was editor of *Organic Gardening* for 12 years and currently is senior editor of *National Gardening* magazine.

WARREN SCHULTZ is guest editor of this handbook, and contributed the sections on ants, billbugs, cherry fruit flies, chinch bugs, tarnished plant bugs, thrips and tomato hornworms. He is the author of *The Chemical-Free Lawn* and a member of the board of directors of the National Coalition Against the Misuse of Pesticides. He is former editor in chief of *National Gardening* magazine and senior editor of *Organic Gardening*.

EMILY STETSON contributed the sections on corn earworms, cutworms, European corn borers and fall armyworms. She is a freelance writer based in Charlotte, Vermont, and the former managing editor of *National Gardening* magazine.

SHELLY STILES contributed the sections on boxelder bugs, bronze birch borers, evergreen bagworms, gypsy moths and tent caterpillars. She is program manager of Bennington County Conservation District and writes from her home in the Taconic mountains.

JAMES E. ZABLOTNY contributed the sections on aphids, darkwinged fungus gnats, mealybugs, scales, spider mites and whiteflies. He is a doctoral student in the Department of Entomology at Michigan State University. He helps tend the botany department's orchid collection and practices biological control whenever possible.

Aphid

Leafminer damage on columbine

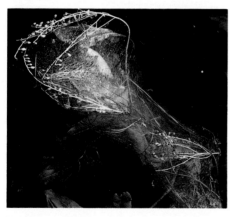

Tent caterpillar

Whitefly

Scale with ants seeking honeydew

Mealybug

INTRODUCTION

BY WARREN SCHULTZ

Having grown up on a traditional family farm in the 1950s, I have at least one fond memory involving pesticides. There would come a day every spring when my Dad and my uncle would prepare a cutworm bait concoction. We'd all gather in a dusty room in the barn, around a big zinc tub, and mix box after box of brown sugar with some mysterious poison.

My brothers and I would hover around, waiting to dip into the boxes of brown sugar — before it was laced with pesticide. But I still don't know what poison we were exposed to.

Pesticides still had an innocent aura back then. They were still pretty new to the scene, and thought of as miracle cures. Better living through science came to the farm. It wasn't until 1944, for example, that herbicides became an option with the introduction of 2,4-D. And about the same time, pesticides like the late but not lamented chlorinated hydrocarbons DDT, aldrin and dieldrin were taking the agricultural world by storm

Up until then, most pest control was natural pest control. Natural control has been in practice, of course, since the first farmer squashed a beetle or caterpillar. It was legitimized in the 1880s when the Vedalia ladybug was imported en masse to eradicate scale and mealybugs in California.

Even as recently as the 1950s, textbooks on pest control began with instructions for natural controls: predators, handpicking, crop rotation, traps. Plenty of space was given to sprays such as rotenone, pyrethrum and horticultural oils.

But the times they were a changing. New chemical pesticides flooded the market. In the rush to embrace synthetic chemical pesticides, many reliable natural controls were dropped like an old pair of shoes. Research into natural controls halted at universities across the country. The use of predators and parasites slowed to a crawl because the heavy use of broad-spectrum pesticides killed off beneficial insects as well as the pests.

In the U.S., pesticide use increased tenfold from 1945 to 1989, but all those chemicals didn't help to win the "war" against bugs. During that period, crop losses from insect damage actually increased from 7 to 13 percent. That's partly because pests have been able to stay ahead of the pesticides. More than 500 species have developed resistance to pesticides over the past several decades.

Today there are more than 600 active pesticide ingredients registered in the U.S. And about 10 percent of them are known or suspected carcinogens.

As pesticides proliferated, an awareness of their dangers grew. Rachel Carson got people's attention with the publication of *Silent Spring* in 1962. In December of 1970, the Environmental Protection Agency was created, and registration of pesticides was handed over to it.

In the 1970s there was a grassroots revival of organic gardening practices. The country saw a surge of "home-brewing" and wishful thinking as pest-control measures. Books and magazines extolled the efficacy of homemade sprays of garlic, onion, marigolds and ground-up bugs. Companion planting was all the rage. It was commonly believed (and still is by some) that certain plants could "protect" other plants from insect attack. For example, it was said that tomatoes or mint planted near cabbage would repel the cabbage maggot. Unfortunately, these funky controls were unreliable at best. And natural gardening got a bad name as a fringe-group activity.

But all the publicity sparked the public's interest. And as more evidence came in that pesticides were poisoning food and environment, the market began to respond to the increasing clamor for safer measures. The agriculture establishment actually began borrowing ideas from organic gardeners and looking to the past to revitalize some of the old techniques.

Integrated pest management was born. Farmers began cutting back on routine spraying of pesticides, monitoring to pinpoint target pests and looking for alternative means of control. Researchers turned their attention back to predators and microbial controls. People began digging back into old files and old textbooks. In some cases they found that so-called new methods were not new at all. Many were surprised to see, for example, that the "new" super-refined horticultural oils, the so-called summer oils that can be used as insecticides year-round, were recommended back in the 50s, but then neglected in favor of stronger synthetics.

Corn earworm

Bean weevil damage to fava bean

Snail on cabbage leaf

Parasitized aphids on carrot

Leafminer damage to chard

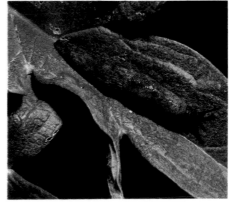

Slug

A clean, healthy greenhouse where beneficial insects are used to control pests

The Brooklyn Botanic Garden has a solid history in natural gardening. In fact, this isn't the first time that this handbook series has looked at natural pest control. There have been at least two handbooks, one in 1960 and one in 1975, devoted to the subject. They've included articles on parasites and predators, natural sprays and even companion planting.

In these pages you'll find descriptions of 52 common pests and the latest natural techniques for controlling them. You'll learn that it's true what organic gardeners have said all along — pick a pest, and you can almost surely find a natural control for it.

HOW TO USE THIS BOOK

This book is designed to make natural pest control as easy as possible. Each entry in the Encyclopedia of Pests that follows includes a description of the insect and its life cycle. (See the next page for a brief explanation of the stages through which insects typically develop. It's important to be familiar with these stages because you won't always see the adult, and many times it's not the adult that does the damage.) Also included in each entry are the most effective controls. Only natural controls are covered, including growing tips, row covers and other "physical controls," natural predators and natural pesticides, which don't persist in the environment and thus wreak a minimum of ecological havoc.

Let's say you're out in the vegetable garden and you notice that your bean leaves are being chewed to bits. What do you do? First, examine the plant carefully for signs of the pest. If you can find the offending insect, you're in luck — check each pest entry listed in the index under "beans," turn to the relevant page in the Encyclopedia of Pests and look at the color illustrations of the insects known to attack beans to find the one that's decimating your plants. If you can't find the pest itself, note the type of damage that it has done. For example, some pests skeletonize the leaves of the plant, while others suck the sap and cause the leaves to begin to yellow and curl. Check the section labeled "Symptoms" in each pest entry until you find a description that most closely characterizes the damage done to your plants.

Once you've identified the pest, check under "Controls" for tips on how to stymie or eliminate it. For detailed information on cultural and physical controls, see page 76. The use of natural enemies to control pests is discussed in "The Predator Patrol," beginning on page 86. For more on natural pesticides, including horticultural oils, insecticidal soaps, microbial pesticides and botanical pesticides, see page 95. Every effort has been made to provide complete and up-to-date pest-management information. However, changes in pesticide regulations occur constantly as more data on the safety of these products becomes available. That's one reason why reading product labels carefully before you buy or apply any pesticide is so important.

THE LIFE CYCLES OF INSECTS

Insects undergo changes in form as they develop into adults. Scientists call this process metamorphosis. The forms that different species take vary, but there are two basic kinds of metamorphosis: simple and complete.

Among the insects that experience simple metamorphosis are aphids, leafhoppers, earwigs, thrips and scale. After hatching, the immature stages of these insects, called nymphs, periodically shed their skins, a process known as molting. During the final molt they develop into the adult stage. Both nymphs and adults resemble each other and have similar feeding habits.

Moths such as that of the cabbageworm illustrated below, beetles, flies and lacewings are some of the insects that undergo complete metamorphosis. After they hatch, they enter the larval stage. It is the larvae of these insects that often cause the most injury to plants. The larvae molt repeatedly. When the larvae are fully developed, they enter the next stage of development, the pupa. Insects in the pupal stage do not feed but undergo dramatic changes in form and finally molt for the last time and become adults. Larvae and adults usually look very different and have very different feeding habits.

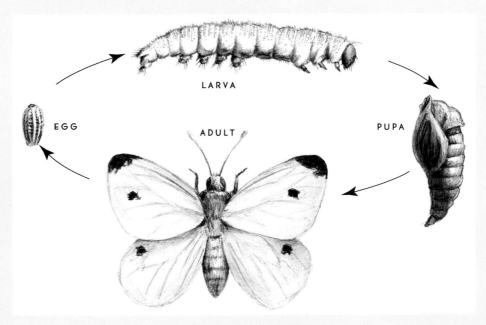

LARVA

EGG

ADULT

PUPA

ENCYCLOPEDIA OF PESTS

ANTS

Ants are irrepressible. It's virtually impossible to annihilate them. Even slowing them down is a major challenge.

They're ubiquitous. Ants are everywhere across North America. There are more than one dozen common, troublesome species, including the little black ant, the odorous house ant, the cornfield ant and the imported fire ant.

Ants are omnivorous. They eat just about anything from cookie crumbs to hamburger to aphid honeydew to garden seeds to plant leaves. Some species are even beneficial, as they feed on more harmful insects.

However, many species are serious pests to humans because they bite and sting. Take the imported fire ant, which lives in mounds up to 2' tall. When those mounds are disturbed, the ants come swarming out, ready to attack. When they encounter flesh, human or otherwise, they go into a frenzy of biting and stinging. Each ant grabs a piece of skin, pulls it up with its pincers and stings repeatedly. On humans, each sting results in a burning, itching blister. Other ants are not so vicious, but many species do bite or sting.

LIFE CYCLE Ants all live in colonies, led by one or more queens. The life cycle of ants varies considerably from species to species. In general, the queen lays eggs from which larvae hatch, usually within 30 days. The larvae develop into a pupal stage 10 to 60 days later, and emerge as adults, usually within 4 weeks.

15

Destroy the queen and the colony will die or disperse. Perhaps the best way to destroy the queen is with boiling water. Make sure you have enough water to penetrate to the deepest part of the anthill, as that's where the queen resides. Wait until late in the day, when most ants have returned from foraging, to flood the ant hills.

Other natural means of ant control: Outdoors, spray with insecticidal soap or citrus oil. Some botanical pesticide products have been registered by the Environmental Protection Agency for use against ants. Check the product labels for information on which pests they control and how to use them safely.

Indoors, remove sources of food. Keep counters and floors clean. Plug up cracks used as entryways by ants. Store food in pest-proof containers. Use pyrethrum/silica gel or boric acid baits. Spray with soapy water.

APHIDS

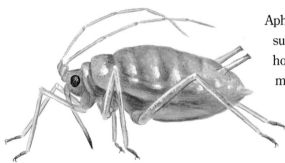

Aphids are small, pear-shaped insects that suck sap from the stems, leaves or roots of host plants. Typical species found on ornamentals or house plants are yellow-green and range from 1/16" to 5/16" in length. Aphids feed on a variety of plants, including impatiens, chrysanthemums, orchids, cyclamen and many other herbaceous plants, vegetables, trees and shrubs. Aphids can also transmit plant viruses to healthy plants during the feeding process.

Typical species overwinter in the egg stage. Hatching occurs in spring. Spring hatchlings are female and when mature give birth to many live young. Winged forms, which develop as the population on the host plant increases, disperse widely to seek out more host plants. During the growing season aphids reproduce asexually, producing multiple generations. Toward the end of the season, males and females mate, producing eggs for the following year. Thus aphid populations grow quickly because of the many asexual generations and are able to successfully overwinter in the single bisexual generation.

SYMPTOMS Clusters of feeding aphids cause stunting and deformation of leaves and stems. Like other sap-feeding insects, aphids can produce copious amounts of honeydew, a sticky, sugary liquid. Ants, which feed on the honeydew, are often another sign of an aphid infestation. Sooty mold grows readily on honeydew deposits and can cover foliage quickly, reducing photosynthetic efficiency.

CONTROLS Many aphids can be destroyed with jets of water or with a cotton swab impregnated with rubbing alcohol. About 4 teaspoons of a mild dishwashing detergent in a quart of water may also be effective.

In nature aphids have many natural enemies that hold populations in check, including syrphid flies, green lacewings, lady beetles and certain species of gall midge. You can attract many aphid parasites by growing flowering plants, which produce nectar that the predators need as an energy source. Larvae of the predator gall midge (*Aphidoletes* species) are sold through some outlets for controlling aphids in commercial greenhouses. Native braconid and chalcidoid wasps parasitize aphids, but are much slower acting than the predators.

APPLE MAGGOT

A native insect widely distributed throughout most of the major apple-producing regions of North America, the apple maggot (*Rhagoletis pomonella*) does the most serious damage in the Northeast and eastern Canada. Hawthorn, apricot, nectarine, plum, pear and cherry, as well as apple and crab apple, serve as hosts.

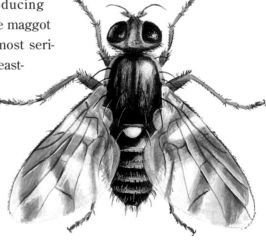

The adult fly is about 1/3" in length, with dark markings on the wings. The larva is a pale, legless maggot without a distinct head, also about 1/3" long.

The apple maggot spends the winter as a pupa, buried shallowly in the soil around the bases of previously infested trees. Adult flies emerge in late June and July. During the first days after emergence, the flies mate and feed. After 10 to 14 days, female flies begin to search out fruit and insert eggs under the skin.

Eggs hatch in about a week and the young maggots feed and burrow through the fruit, spending 3 or 4 weeks there. When fruit are heavily infested, they may drop; full-grown larvae exit the fruit and burrow into the soil, where they pupate. There is one generation per year in most of the U.S.; a small, second generation can occur in the more southern areas.

Damage is caused primarily by the tunneling of the larvae through the apple and other susceptible fruit. The fruit then turns dark brown due to bacteria carried by the maggots. Adult flies also can produce visible damage as they puncture the developing fruit in the process of egg laying, giving the maturing fruit a dimpled appearance.

Apple maggot adults can be monitored and, in some cases, controlled with colored sticky traps. Yellow sticky boards are particularly attractive to younger flies. Older flies, seeking fruit for egg laying, are more readily trapped with red sticky spheres, which look like apples. Where apple maggot problems are moderate, enough flies can be trapped with a few traps per tree to provide adequate control. Traps are most efficient when they are placed so that sunlight directly strikes them and if leaves and fruit are removed from the area immediately around them.

Good sanitation is important. Pick up and remove dropped fruit promptly to eliminate larvae that have not yet emerged.

Because honeydew, excreted by aphids and some other sap-sucking insects, is an important food source for newly emerged adults, take steps to control aphids on and around fruit trees during midsummer.

Some botanical pesticide products have been registered by the Environmental Protection Agency for use against apple maggots. Check the product labels for information on which pests they control and how to use them safely. Sticky traps are also very useful for indicating when flies are active so that insecticides can be applied at the appropriate time.

BILLBUGS

Billbugs are weevils that feed on lawn grasses. The adult billbug is named for its most prominent feature, a long snout that ends in a set of mandibles. But it's the larvae that do most of the damage.

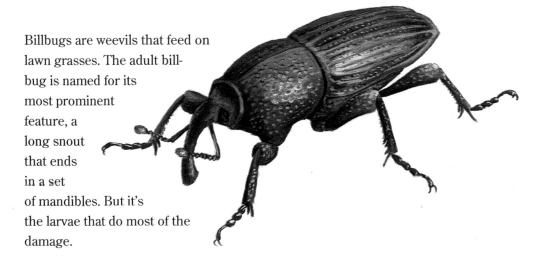

LIFE CYCLE The gray or brown adults are 1/4" to 1/2" long. They overwinter as adults throughout the U.S. and emerge in late spring to lay eggs on the stems of grass plants. Small, legless, white larvae soon emerge from the eggs and feed on stem tissue, causing the shoots to turn brown and then die. As the weather warms up, the grubs burrow into the soil and begin feeding on grass roots.

SYMPTOMS Lawns that are infested will appear brown and drought stressed in midsummer. You may also be able to spot the adults on driveways and sidewalks earlier in the spring.

CONTROLS Build up your lawn's resistance by aerating, watering deeply and top-dressing with organic matter. If you're reseeding, use a billbug-resistant variety.

If you find one adult or 10 larvae per square foot of lawn, you may want to consider using an insecticide. Some botanical pesticide products have been registered by the Environmental Protection Agency for use against billbugs. Check the product labels for information on which pests they control and how to use them safely.

BOXELDER BUG

The boxelder bug (*Leptocoris trivittatus*) is found from the eastern U.S. and Canada west to Nevada feeding on leaves and seeds of boxelder and, occasionally, silver maple.

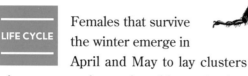

LIFE CYCLE Females that survive the winter emerge in April and May to lay clusters of eggs on and near boxelder and other hosts. In about two weeks, the eggs hatch into the first of several nymph stages, all red-bodied with black proto-wings. Adults are about 1/2" long with black bodies and red-orange chevron-shaped markings.

SYMPTOMS Plant damage is usually insignificant. The bug's habits around the home, however, can be a considerable annoyance.

Boxelder bugs are usually solitary. But in autumn (October in the southern part of their range), females become gregarious. They often gather by the thousands, swarming on the walls of buildings and on objects in the yard, seeking entry under the eaves, inside the attic and the walls of houses, garages and outbuildings where they spend the cold months together. The swarming period is often unpleasant for homeowners, as are winter warm spells, when the bugs may temporarily break dormancy and invade the family's living spaces.

CONTROLS Remove female boxelders. Sweep up and destroy swarming adult bugs.

Some botanical pesticide products have been registered by the Environmental Protection Agency for use against boxelder bugs. Check the product labels for information on which pests they control and how to use them safely.

BRONZE BIRCH BORER

Found throughout the northeastern and north-central U.S., the bronze birch borer (*Agrilus anxius*) is our worst pest of both native and many introduced white-barked birches, including, gray birch and European white birch and its cultivars. Stressed trees are more likely to be attacked.

LIFE CYCLE All life stages of this pest are elusive, but the adult can sometimes be seen feeding on birch foliage. Look for a 1"-long insect that looks something like a lightning bug. It has a sooty green-black body and a coppery or bronze head. (If captured, it may audibly vibrate. It and most other "buzz beetles" are important pests of North American forests.) Adults emerge to feed and mate in early to midsummer in the northeastern states.

Eggs are laid until late August under exfoliating bark or in cracks in trunks, where they soon hatch into yellow-white, flat-headed larvae up to 1" long. Although adult feeding does little damage, larval feeding is often deadly.

Bronze birch borer larvae spend their entire cycle — months in the southern part of their range, a year or more to the North — inside the tree, chewing their way through the tissues that conduct water and nutrients throughout the plant.

SYMPTOMS Yellowing leaves, witches' brooming and dieback of stems within the clump are common and visible signs of the presence of the bronze birch borer.

CONTROLS Water and feed as needed to prevent stressing of the tree. Remove dying or dead stems.

Replace with a resistant species, such as *Betula lenta* (black birch) and *B. nigra* (river birch).

CABBAGE LOOPER

The cabbage looper (*Trichoplusia ni*) is found throughout the U.S. Cabbage looper caterpillars are light green with white back stripes and 1-1/2" long. The name "looper" comes from its U-shaped crawling habit. The night-flying adult cabbage looper moth is 1-1/2" wide and a mottled gray-brown. There is a distinct small, silvery spot on each forewing. Moths appear in early spring in the South, and mid to late June in the North.

LIFE CYCLE

Each female places about 300 pinhead-size, greenish white eggs on the upper surfaces of the leaves. Caterpillars, which start out with brown heads, emerge in about 4 days. Within 2 weeks, each caterpillar weaves an almost transparent white cocoon in which to pupate. The cocoons are attached to the undersides of host plant leaves. In two weeks, another generation of moths emerges.

In cold-winter areas, there can be 2 to 3 generations per growing season. The caterpillars overwinter wrapped in green-brown cocoons attached to host plants. In warm-winter areas there can be 7 generations; in some areas, year-round activity is possible.

SYMPTOMS

If irregularly shaped or rounded holes begin appearing on the older leaves of your cabbage, lettuce, spinach, potatoes, tomatoes or celery, keep a lookout for cabbage loopers. Other clues are greenish brown waste pellets on the undersides of damaged leaves and, of course, the caterpillars themselves.

CONTROLS Plant resistant cabbage varieties. Clean and dispose of all crop remnants in the fall. Cover your plants with synthetic row covers to keep the adult moths from laying eggs.

Parasitic wasps can be effective against caterpillars or eggs. *Bacillus thuringiensis* is also effective. It is usually best to use *Bt* when caterpillars first hatch.

Some botanical pesticide products have been registered by the Environmental Protection Agency for use against the cabbage looper. Check the product labels for information on which pests they control and how to use them safely.

CABBAGE MAGGOT

This root maggot attacks mostly brassica-family vegetables such as cabbage, cauliflower, radish, broccoli, brussels sprouts and turnip. The cabbage maggot (*Delia radicum*) is a northern insect pest that is a problem from the east to west coasts north of a line drawn from North Carolina through Illinois into northern California. It wreaks its damage as far north as southern Canada.

LIFE CYCLE The ash-colored, 1/4"-long adult fly emerges from the soil and lays small white eggs in April and May within 2" of the plant's stem. The eggs hatch in a few days and the white, tapered, 1/3"-long, pointy-headed larvae feed on the host plant's roots for one month, pupate in the soil, then start another generation. There are usually 2 to 4 generations a year, depending on the locality.

SYMPTOMS The larvae favor cool, wet weather. Their feeding weakens the roots, eventually causing the lower leaves to yellow and the plant to wilt and die. Root crops are often scarred with brown-streaked tunnels, where diseases such as black rot often develop.

CONTROLS Physical barriers that prevent the adult from laying eggs are the best controls. Screen transplants in spring and fall with a cone or floating row cover or place 3"-diameter tar paper collars on the soil around the base of the transplant.

Diatomaceous earth granules sprinkled around the base of transplants can kill the hatching larvae, but repeated applications may be needed.

CABBAGEWORM

As the name suggests, the cabbageworm (*Pieris rapae*) prefers chewing on members of the cabbage family: broccoli, cabbage, brussels sprouts and cauliflower. These imported pests are found throughout the U.S., almost always in conjunction with the cabbage looper.

The slow-moving cabbageworms are almost the same pale velvet-green color as a cabbage leaf, although there is a slender yellow back stripe and a broken stripe along each side. To find them, look for messy, dark-green waste matter. The cabbageworms are usually feeding on leaf undersides right above this.

LIFE CYCLE After a 3-week caterpillar stage, transformation into butterfly form occurs within a greenish chrysalis. This is usually suspended from a host plant by a silken belt.

Two-inch-wide, bright white butterflies with black wing spots emerge within 2 weeks. Females flit down regularly, each gluing up to 100 yellowish eggs on the undersides of host-plant leaves. Larvae, pale green with a faint yellow back stripe, hatch in 4 to 8 days. Within 3 weeks they are mature, measuring 1-1/4" long.

There can be 3 broods a year in the North, and up to 6 broods in the South, where butterflies may remain active all winter. During cold weather they hibernate under old stalks and weeds. They are among the first butterflies to emerge in early spring, often seeming to play twosome tag.

 SYMPTOMS Cabbageworms chew holes in the leaves of host plants, beginning with the outside leaves and progressing inward.

 CONTROLS Handpick and destroy caterpillars. Clean up after harvest to destroy overwintering cover. Plant resistant cabbage varieties. Protect young plants with floating row covers.

Tachinid flies parasitize cabbageworm larvae. Trichogramma wasps parasitize the eggs. You can attract tachinid flies and other beneficial organisms by growing herbs and flowers in or near your vegetable patch; the wasps are available from mail-order suppliers. *Bt* and insecticidal soaps are also effective.

Some botanical pesticide products have been registered by the Environmental Protection Agency for use against cabbageworms. Check the product labels for information on which pests they control and how to use them safely.

CARROT RUST FLY

One of many native European insects that stowed away to North America, the rust fly (*Psila rosae*) attacks not only carrots but other carrot-family plants as well, such as celery, dill, parsnip, parsley, coriander and fennel. It is mostly a problem in the northeast and northwest U.S. and southern Canada.

 LIFE CYCLE The shiny black, 1/4"-long adult fly emerges from the soil, after overwintering on plant debris, in early May and lays white eggs at the bases of your plants. The eggs hatch in a few days and the 1/3"-long, yellowish white larvae tunnel into the roots, leaving behind rust-red colored excrement (hence the name). They feed for 3 to 4 weeks, pupate and start another generation in fall. It's the second generation that does the most damage on carrot roots. There may be a late third generation in warmer areas. Carrots are still edible if the damage is minor, but severely tunneled roots may develop secondary disease infection and rot before harvest.

SYMPTOMS Carrot tops sometimes yellow. Holes, scarring and tunneling may be evident in the root, and secondary rot may also occur.

CONTROLS The best controls are removing any carrot-family plant debris in fall, where the adult can overwinter, and planting carrots under floating row covers to prevent the adult from laying eggs. Removing carrot-family weeds such as Queen Anne's lace from the area and tilling the soil in fall to expose overwintering pupae to winter weather and predators also will reduce the population.

CHERRY FRUIT FLY

The cherry fruit fly (*Rhagoletis cingulata*) is a native American insect whose maggot lives in the flesh of cherries, causing them to rot. It is related to the apple maggot, which it resembles.

The fly is a bit smaller than the apple maggot. It is about 1/5" long with wings about 1/2" across. The body is black, the head and legs are yellow to brown. The sides of the thorax are marked with a yellow band, and the wings are crossed by 4 black bands. The maggot is white and about 1/4" long.

LIFE CYCLE The flies overwinter as pupae in soil near cherry trees. The adults emerge in late spring and lay eggs in the developing fruit.

The eggs hatch in the fruit, and the larvae penetrate to the pits, feeding on the flesh and causing the cavity to rot. When full grown, the larvae emerge from the fruit and pupate in soil or debris at the foot of the tree.

SYMPTOMS There is often no evidence of infestation unless the cherry is cut open to reveal either a maggot or its tunnel.

CONTROLS Removing and destroying all fallen fruit and debris helps to break the life cycle, as does deep cultivation of surrounding soil in early spring. The adults may be trapped with yellow sticky traps hung in the trees. More complete control can be attained by covering the trees with floating row covers after the fruit has formed.

Some botanical pesticide products have been registered by the Environmental Protection Agency for use against the cherry fruit fly. Check the product labels for information on which pests they control and how to use them safely.

CHINCH BUG

The chinch bug (*Blissus leucopterus leucopterus*) is a common turf pest across the North American continent. The adult is about 1/5" long, with a black body. Its white wings lie folded over each other on the abdomen, and are marked by a small black triangle on their outer margins. The bases of the antennae and the legs are red. The younger bugs are yellow or red, becoming darker as they grow older.

During the winter, the bugs hibernate under plant debris. Females lay small yellow-white eggs in spring from April to June. It is the nymphs that do the most damage to turf, feeding particularly on Kentucky bluegrass, fine fescues, bentgrass, St. Augustine grass and zoysia.

When they first hatch, the nymphs are smaller than a pinhead, and are bright red with a white band across the back. They feed on stems and leaves of grass, and continue to feed as they grow. They inject a salivary fluid into the grass, causing it to turn yellow and eventually die in patches.

Chinch bugs have an odd, offensive odor, especially when crushed. A severely infested lawn will have a detectable odor. To check a lawn for a chinch bug infestation, cut both ends from a 1-pound coffee can and press it firmly into the turf. Fill the can with water. If your lawn has chinch bugs, they will float to the surface. Fewer than 20 in the can is not a serious infestation. If there are more than 20 it's time to treat the lawn.

There are several cultural controls that will discourage chinch bug infestations. Because they prefer hot, dry conditions, you can discourage them by irrigating well. Cutting back on nitrogen fertilizer also makes the grass less succulent and appealing. Be sure to seed with grass varieties resistant to chinch bug damage.

CODLING MOTH

Codling moth *(Cydia pomonella)* is considered the most destructive insect of apples and pears in most of the U.S. Larvae of the codling moth are usually *the* worm in a wormy apple. The larvae tunnel through the fruit, ultimately feeding on the developing seeds in the core.

Codling moth overwinters in the pupal stage. The adult moths typically emerge to lay eggs shortly after petal fall. The caterpillars of the first generation feed on the leaves, then move to the developing fruit as it becomes available. These become full grown, pupate and turn into moths that produce the second — and most damaging — generation in July. At this time eggs are laid on the surface of the fruit into which the young caterpillars tunnel.

Two, and sometimes 3, generations occur per year. The midsummer (second) generation is most damaging to apples and pears.

Conspicuous crumbly brown excrement is visible where the insect entered the fruit. Cutting the fruit open exposes the tunneling to the core.

Codling moth larvae seek protective sites, such as bark flaps, to pupate; removing loose bark helps inhibit pupation. Alternatively, providing a favorable pupation area, such as a trunk band of corrugated cardboard or burlap, can encourage pupae to congregate; you can then collect and destroy them. Collect and change the traps every 7 to 10 days, so that the pupae do not emerge as adult moths and escape.

Adult moths of both sexes can be attracted to traps baited with dilute molasses or molasses and beer. Put this fermenting mixture in cut-out milk jugs or similar containers and hang them on the tree.

Developing fruit that are touching are particularly susceptible to infestation by codling moth larvae, so be sure to thin the fruit.

A great many parasitic wasps are known to attack codling moth. Trichogramma wasps parasitize the eggs and so are best released during periods of egg laying. These are available from mail-order suppliers. Codling moth larvae are susceptible to *Bacillus thuringiensis,* although their habit of feeding within the apple can prevent them from ingesting enough for control.

Codling moth is usually controlled with insecticides applied to cover fruit so that newly emerged caterpillars are killed before they can enter. Pheromone traps are useful for detecting periods of peak moth activity, when insecticide applications are most effective. Some botanical pesticide products have been registered by the Environmental Protection Agency for use against codling moth. Check the product labels for information on which pests they control and how to use them safely.

Sanitation is important. Pick up fruits that fall to the ground.

COLORADO POTATO BEETLE

Adult Colorado potato beetles are about 3/8" long, with black and yellow lengthwise stripes on their wing covers and black dots just behind their heads. Eggs are yellowish orange and laid in clusters on the undersides of leaves. Larvae are plump and shaped like beetles. Young larvae are red but have black heads and legs; older larvae are orange with two rows of black dots down each side of their bodies.

This pest occurs throughout North America. Potatoes as well as tomatoes, eggplants and peppers are affected.

LIFE CYCLE Adult beetles overwinter in undisturbed woody areas. After emerging from the soil in early spring, they move into gardens to feed and lay eggs. These hatch into larvae, which develop into adults. There can be up to three generations per year.

SYMPTOMS Chewed leaves have a lace-like appearance. In severe cases, the plant can be defoliated.

CONTROLS A thick mulch of straw or other organic material will confuse the beetles and keep them from finding vulnerable plants. Handpick adults, larvae and eggs. Surround your garden with a steep-sided trench lined with plastic; as the beetles crawl toward their host plants, they'll fall into the trench and become trapped. Collect them from the trench and destroy them.

Bacillus thuringiensis san diego is effective against the larval stage (not the adult stage). Apply it as soon as you see the larvae, making sure to cover leaves thoroughly, both upper and lower surfaces.

Lady beetles and spined soldier bugs feed on the larvae. Attract lady beetles with pollen and nectar plants; maintain permanent perennial plantings near your garden to harbor spined soldier bugs.

CORN EARWORM

The larvae of the corn earworm (*Helicoverpa zea,* also known as the tomato fruitworm and the cotton bollworm) victimize corn, tomatoes, potatoes, beans, peas, peppers and squash throughout North America. In frost-free regions, the earworm may go through as many as 7 generations per year; in the North there are rarely more than 2 generations.

LIFE CYCLE The greenish gray or brown adult moths lay off-white, ribbed, dome-like eggs singly on corn silks or the undersides of leaves in spring. Hatching occurs a few days later, and the larvae feed on corn silks, leaves and fruit for 3 to 4 weeks, then leave the ear to pupate in the soil. The 1-1/2"-long larva is green, white or light yellow, with dark stripes down either side of its body. Unlike the European corn borer, the earworm larva is cannibalistic, so you'll rarely find more than 1 or 2 per ear of corn.

SYMPTOMS An earworm can quickly disfigure or destroy an ear of corn by spoiling the ear's end, inhibiting full pollination and inviting molds and other diseases. Other victimized crops will show leaf and stem damage.

CONTROLS In the North, timing corn plantings so that the silking occurs between the two peaks of earworm feeding can help avoid an infestation. In more southern areas, where overwintering occurs, aim for a mid-season crop, as early- or late-planted sweet corn usually suffers the greatest damage. If possible, choose varieties with long, tight husks that extend beyond the tips of the ears to reduce injury. Examples are Texas Honey June, Geronimo and Silver Queen. Handpick (pull back the corn tips and remove the earworms) *after* the silks have turned brown, indicating pollination is complete.

Removing or chopping garden debris and fall tilling kill or expose the overwintering pupae to predators and cold, reducing the number that will survive to spring. In extreme southern areas, where earworm development continues uninterrupted year-round, this same technique works well for a different reason: Tilling during winter warm spells forces the emergence of the moths at a time when no food is available, thus reducing later populations.

Both tachinid flies and trichogramma wasps parasitize earworm eggs. You can lure tachinid flies and other native beneficials to your garden by planting herbs and flowering plants near the corn patch. Trichogramma wasps can be purchased from mail-order suppliers.

Bacillus thuringiensis kills the young larvae, but must be applied before they burrow into the ear or fruit. Apply *Bt* granules on newly emerging silks every 3 to 4 days for 2 weeks; on other plants, spray as soon as you detect the larvae. Another time-honored treatment is to suffocate the earworms by placing a dropperful of mineral oil into the tip of each ear when the silks have withered but before they turn brown. Repeat the mineral oil remedy again in a week.

CUCUMBER BEETLES

There are several species of beetle commonly called "cucumber beetles." The spotted cucumber beetle is 1/4" long, yellowish green with 12 black spots and a black head. The striped cucumber beetle is 1/5" long, and yellowish orange with three black stripes down its back. They range throughout most of North America, infesting many vegetables and fruits, including cucumbers, melons, squash and other cucurbits; corn, potatoes, tomatoes, eggplant, beans, peas, beets, asparagus, cabbage and lettuce; plus peaches and other soft fruits.

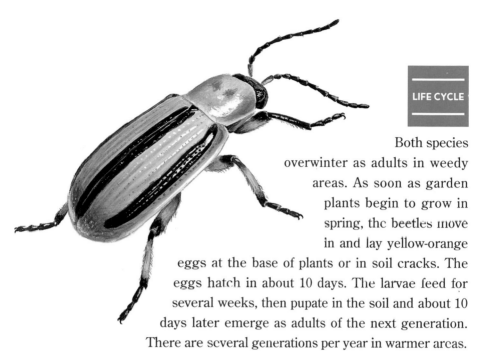

Both species overwinter as adults in weedy areas. As soon as garden plants begin to grow in spring, the beetles move in and lay yellow-orange eggs at the base of plants or in soil cracks. The eggs hatch in about 10 days. The larvae feed for several weeks, then pupate in the soil and about 10 days later emerge as adults of the next generation. There are several generations per year in warmer areas.

SYMPTOMS Damage from chewing by both adults and larvae can be seen on stems and leaves. Adults also chew on blossoms and fruit.

CONTROLS Cucumber beetles are some of the most difficult pests to control organically. Look for cucurbits with resistance to cucumber mosaic virus and bacterial wilt, diseases spread by the beetles as they feed.

Tuck row covers around your cucurbits at seeding time, allowing lots of extra fabric to keep the plants covered as they grow. Lift the row covers for several hours each day for pollination when the plants are in flower.

Several beneficial insects, including tachinid flies, Pennsylvania leather-wings and braconid wasps, parasitize various stages of the cucumber beetle. These are not available commercially but can be lured to your garden if you grow pollen and nectar plants (including herbs such as dill).

CUTWORMS

Cutworms are among the
most destructive garden
pests found throughout
North America, and they
can be one of the most maddening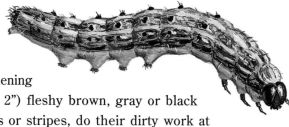
to control. The large (up to 2") fleshy brown, gray or black
caterpillars, often with spots or stripes, do their dirty work at
night, feeding on vegetables (particularly peppers, tomatoes, eggplant, corn, cabbage and beans), flowers, vines and trees.

There are approximately 3,000 species of cutworms in North America, named according to their feeding habits: surface or tunnel-making cutworms (which topple seedlings at or just below the soil surface), climbing cutworms (which attack aboveground parts of the vegetables, herbaceous plants and even fruit trees, sometimes traveling in hordes) and subterranean cutworms (which feed on roots and underground stems of vegetables, grains and grasses). It's possible to have several types of cutworms in one garden, each at a different life stage. These pests can be especially severe in a garden that's been recently converted from lawn or field grass.

LIFE CYCLE Most common cutworms overwinter as larvae, though some species hibernate as adult moths, pupae or eggs. The larvae feed from spring through early summer, hiding in surface litter or the top few inches of soil by day. Then they tunnel into the soil to pupate and transform into adult moths. Within a week after the 1" to 2", mottled gray or brown cutworm moths emerge, the females deposit masses of hundreds of tiny (1/50") white eggs on weeds and leaves. Caterpillars hatch in two days to several weeks. Northern areas usually have only a single generation per year, but in the South there may be as many as 4.

SYMPTOMS By the time you discover the symptoms — well-watered yet wilting plants, severed seedlings or plants with missing leaves, buds and fruits — it's often too late to save the crop.

Remove weeds and crop residues in late summer to discourage the moths from laying eggs, and till the soil to expose the overwintering larvae to natural predators (birds, ground beetles, moles, toads) and harsh weather. Intensive weeding in spring 2 weeks to 10 days before transplanting or seedling emergence will also minimize egg laying and starve any surviving or newly hatched larvae.

Cutworm "collars," made from milk cartons, PVC pipe, metal cans, aluminum foil, tarpaper squares or cardboard tubes and pressed 1" deep into the soil around each plant, are the traditional solutions for seedling protection. Placing a small stick or nail directly next to the seedling's stem is also reputed to deter cutworm damage, as the larva can't wrap itself around the stem to fell it. Though they won't help protect seedlings, sticky barrier traps can help prevent climbing larvae from feeding on fruit, leaves and buds of sturdier plants and trees. Handpicking the cutworms at night with a flashlight may be helpful.

Beneficial insects can help control a persistent cutworm problem. Trichogramma wasps, applied in early spring, will parasitize newly laid cutworm eggs, and beneficial nematodes can be added to the soil in spring or fall to parasitize the larvae. These are available from mail-order suppliers. Native beneficial insects such as tachinid flies and braconid wasps also parasitize cutworms and can be lured to your garden if you grow herbs and flowering plants. *Bacillus thuringiensis* is an effective control on the young larvae of the surface or climbing species, but must be applied when the cutworms are active.

Homemade baits have little scientific evidence to back them up, but may be worth a try. Cutworms are reputedly attracted to cornmeal, then die from their inability to digest it. A mixture of molasses, wheat bran and sawdust, glopped in a circle around the cutworm's favorite plants, traps the larvae before they reach the stem.

EARWIGS

Earwigs have very short wings and large posterior pincers. The European earwig, a serious pest, is up to 4/5" long, and a dark reddish brown with lighter brown details. Nymphs are similar, but wider for their length and paler brown. Earwigs occur throughout the continental U.S. European earwigs are an important pest in West Coast states, and may also cause serious damage in other parts of the country.

LIFE CYCLE Earwigs feed at night, hiding in the upper soil or crevices in the daytime. Female earwigs lay 20 to 50 eggs in fall or spring. The eggs hatch in spring, and females tend the nymphs until the first molt. There may be a second generation from a June mating. The previous year's adults die by midsummer.

SYMPTOMS All earwigs eat decaying organic matter and some pests, such as insect larvae, aphids and small snails. When European earwig populations are large, they will also feed on any tender plant parts, including seedlings, leaves, soft fruit, cornsilks and flower petals. They eat irregular holes in leaves and flowers and consume cornsilks down to the point where they emerge from the husks.

CONTROLS Eliminate favored hiding places, such as plant debris, pots, ivy or weedy areas.

Use rolled-up-paper traps, beer traps or containers partly filled with vegetable oil. Empty traps daily, shaking the insects into soapy water. Place a sticky band around lower tree trunks.

Insecticidal soap is effective if the insects are hit directly. Diatomaceous earth containing patented baits that attract earwigs is another option. Some botanical pesticide products have been registered by the Environmental Protection Agency for use against earwigs. Check the product labels for information on which pests they control and how to use them safely.

EUROPEAN CORN BORER

The European corn borer (*Ostrinia nubilalis*) is one of the most serious pests of corn in the northern and central U.S. (east of the Rockies and north of Tennessee) and southern Canada. Though corn is the preferred host, tomatoes, eggplant, potatoes, peppers, beans and various flowers (including asters, chrysanthemums, cosmos, dahlias and hollyhocks) may also be affected.

LIFE CYCLE The buff-colored adult moths appear in May and June and begin laying masses of 15 to 35 white eggs on the undersides of leaves. The eggs hatch in 7 days, and the tiny, creamy white, black-headed larvae immediately begin chewing into the corn plants, starting at the leaves, then the whorls and tassels, finally entering the stalks and ears. In a bad infestation, there may be as many as 5 borers in one ear of corn.

The larvae grow to their full size of 1" in about a month, sporting brown spots on each flesh-colored segment, then pupate as reddish brown grubs and emerge as adults to repeat the cycle. There are 1 to 3 generations per year. Cool, rainy weather in early spring reduces the number of borers by inhibiting egg laying and by washing the newly hatched larvae from the plants. Dry summers and very cold winters also keep down the borer population.

SYMPTOMS Shot-like holes in the upper foliage, or holes in stalks or ears signal borer troubles.

CONTROLS Using borer-resistant or -tolerant varieties of corn and timing plantings to avoid the peak periods of borer infestation are the most effective control measures. Avoid very early or very late plantings, and choose early or midseason varieties to avoid the second, more numerous, generation of borers.

Handpicking is tedious, but by slitting the stalks or ears right under obvious borer holes, you may be able to pluck out some of the pests. Guard against future infestations by tilling, chopping or mowing an affected crop — destroying the plants to within 1" of the soil — immediately after harvest to prevent the borers from overwintering in the stalks.

Bacillus thuringiensis will kill the borers, but it must be applied before the larvae tunnel into the stalks. If you suspect a problem, watch for the eggs and apply *Bt* a week later onto whorls, silks and tassels, continuing weekly sprays until the egg-laying stage is past.

Tachinid flies, braconid wasps and lady beetles all prey on the borers. These native beneficials can be lured to your vegetable garden if you grow herbs and flowering plants nearby. Lady beetles are also available for purchase.

EVERGREEN BAGWORM

The evergreen bagworm (*Thyridopteryx ephemeraeformis*), a moth, spends most of its life in a homely cloak. It feeds primarily on arborvitae but also on elms, maples, buckeyes and other shade trees from New York and New England to Florida and west to the eastern edge of the Great Plains.

Few gardeners ever see the adult bagworm moth because only the inconspicuous males fly. (Look for a tufted black body and smoky white or transparent wings about 1" across.) The adult stage of the female, a slug-like, wingless, legless and blind organism, is spent inside the twig- or leaf-covered case for which the pest is named.

LIFE CYCLE Males join females in the case in September or October to mate and lay eggs. Hatching occurs in May or June (in the North). The developing larva, a 1"-long, brown caterpillar, leaves its birth chamber to spin its own portable home, then spends the next 4 or 5 months (or longer in the South) feeding on needles or leaves from within it. At the end of the larval phase, the moth pupates inside the bag, developing into an adult in about 2 weeks. This is an unusually restricted life cycle for members of the moth family, and explains why bagworm infestations don't spread quickly. It also explains why individual trees can be overcome by large populations unless larval feeding is kept in check.

SYMPTOMS Damaged needles turn brown and may persist on the tree for some time. When infestations are heavy, large numbers of brown larval cases hang from twigs and branches.

CONTROLS Hand-pick cases. Spray with *Bacillus thuringiensis berliner* strains.

FALL ARMYWORM

The fall armyworm (*Spodoptera frugiperda*, a type of cutworm) is primarily a southern pest, but migrates to northern areas of the U.S. and to Canada in late summer and fall, often traveling in large groups in search of food. Although best known by gardeners as a pest of sweet corn, the larvae also devour grasses, cotton, alfalfa, peanuts and tobacco, and will attack most garden crops if given the opportunity.

LIFE CYCLE In warm-weather areas, where the fall armyworm is most prevalent and survives winter, the gray, mottled 1-3/4"-long adult moths emerge in spring to lay masses of 100 or more light gray, fuzzy eggs on the leaves and blades of victim plants at night. Larvae hatch in 2 to 10 days, then molt 6 times until they reach their full size of 1-1/2" to 2". The mature larva is green to brown, with a prominent Y- or V-shaped white marking on its head. After about 20 days of feeding and growing, the larvae enter the soil to pupate. Adult moths emerge after 10 days. Northern areas may see 1 or 2 generations per year, while in the South there may be as many as 6. Freezing temperatures kill all life stages.

SYMPTOMS On corn, the larvae bore inside the stalks and ears; on cole crops and other vegetables, the nocturnal feeders devour leaves, heads, stalks, buds and fruit.

CONTROLS Removing garden debris promptly and protecting early crops with floating row covers can help prevent a spring infestation of the egg-laying moths. Handpicking the larvae offers some control, and should the armyworms become too numerous, a spray of *Bacillus thuringiensis* is very effective. Spray both tops and bottoms of the leaves, repeating the treatment every 7 to 10 days as necessary. Beneficial nematodes also parasitize the young larvae and can be purchased from mail-order suppliers.

Trichogramma pretiosum is a commercially available wasp that parasitizes the egg stage of the armyworm. Other natural predators of the larvae include ichneumon wasps, braconid wasps, tachinid flies, ground beetles, birds and toads. Providing alternate host and shelter crops such as clover, alfalfa and flowering herbs for these natural enemies may be the best bet for long-term control.

Some botanical pesticide products have been registered by the Environmental Protection Agency for use against armyworms. Check the product labels for information on which pests they control and how to use them safely. Also keep on the lookout for a new insect growth regulator, Mimic, which has recently been developed to control armyworms and other caterpillars. Like *Bt*, it's specific to Lepidoptera insects, doesn't harm beneficials and is safe for mammals, humans and the environment. The synthesized hormone disrupts the insect's molting process so that the larva forms a new skin inside the old, then starves to death in its "shell."

FLEA BEETLES

There are a number of species but all adult flea beetles are tiny (about 1/10"
long). Some are black or blue-black; others are brownish black with faint stripes
or markings. They jump quickly when
disturbed, like fleas. Larvae measure up
to 3/4" long, and are thin and white
with brown heads. Eggs are white
and nearly microscopic.

Flea beetles occur throughout
North America. Many vegetables,
especially brassicas, eggplants, potatoes and tomatoes, are affected.

 LIFE CYCLE Adults overwinter close to the surface of the soil and emerge in
spring as the plants they feed on sprout. Soon the adults lay eggs in
the soil at the base of plants. After hatching, larvae burrow into the
soil and feed on roots, then pupate and emerge as adults. There are 2 to 4 generations per year.

 SYMPTOMS Leaves damaged by flea beetles look as if they've been shot through
with many tiny bullets.

 CONTROLS Make sure your transplants are big and sturdy before you plant
them out in the garden; larger plants have a better chance of surviving flea beetle attack. Protect young plants with row covers. Give
your garden a noontime shower; flea beetles feed at the height of the day, and
they don't like to get wet. Plant susceptible crops next to tall crops that will
shade them at midday. Mix up plants to confuse beetles. Cultivate the soil frequently to destroy eggs and larvae in the soil.

Attract beneficial native braconid wasps and tachinid flies with pollen and
nectar plants. Apply commercially available beneficial nematodes when flea beetles are in their soil-dwelling stage, or several times throughout the summer.

FUNGUS GNATS

Fungus gnats can be found throughout the year in greenhouses or around potted houseplants. Most species are harmless, but several are a nuisance when population levels are high.

LIFE CYCLE Adult fungus gnats are small (1/8" to 1/4" in length) dark-colored, long-legged flies. Females mate and begin to lay eggs several hours after emerging from the pupa. Eggs are laid on or in the top layer of soil. After hatching, the larvae or maggots begin to feed. Most fungus gnats feed on soil fungi, algae and rotting vegetation, but certain species will feed on the roots or root hairs of cultivated plants. Maggot feeding introduces bacterial or fungal rots, which can kill the plants. The legless maggots are transparent with a dark head capsule.

SYMPTOMS Symptoms of fungus gnat infestation include wilted leaves and stems. The rooted end of the stem may contain rotted tissue and maggots. Stressed plants or new cuttings are usually more susceptible than healthy, mature plants. Ferns and poinsettias are especially prone to fungus gnat feeding.

CONTROLS Adult fungus gnats are easily trapped on yellow sticky cards placed at ground level around infested plantings. Removing all organic debris from around the growing area should reduce food sources. Larvae can be removed during repotting. Be sure to remove any rotten roots, leaves and stems from the plant along with any old soil from the root ball. Nonsterile compost can harbor large populations of maggots. Steam sterilization of potting soil and compost prior to use can kill eggs, maggots and pupae. Keep sterilized potting soil dry in a sealed, heavy plastic bag or container to prevent further contamination.

The beneficial nematode *Steinernema feltiae* is an excellent biological control for fungus gnats. Apply them when watering potted plants and to the greenhouse floor for more thorough control. Some growers report that predaceous mites used for two-spotted spider mite control appear to reduce fungus gnat populations as well.

GYPSY MOTH

The gypsy moth (*Lymantria dispar*) was introduced to the U.S. in 1858. Free of its native European predators, its populations immediately exploded. Now, despite quarantine programs and the importation of several control organisms, it has spread throughout the mid-Atlantic states and southern New England, and has been reported across much of the country. The larvae do extreme damage to oaks and also feed on other tree species.

LIFE CYCLE Larvae hatch in April or early May and feed for approximately 7 weeks before pupating. Adult moths emerge in mid-July. Female adults are flightless, but males are energetic, erratic fliers. After mating and laying eggs, adults quickly die. The remainder of the moth's annual cycle is spent in the egg phase.

So common is the moth that each of its 4 life stages is likely to be seen at some time by the observant gardener. The egg masses, conspicuous at 1" or more in length and clothed in drab white silken hairs as well as the ribbed, shiny brown, empty pupal cylinders, can be seen throughout the year on trees, buildings and other objects. Both moths have a wingspan of about 2" and are patterned with wavy shades of brown or gray, on males on a dusky brown background, and on females on sooty white wings. The bristly larvae are easily distinguished from other caterpillars by the patterning on their backs — 5 pairs of blue dots followed by 6 pairs of red.

 SYMPTOMS Leaves on affected trees are often consumed to the mid rib. Trees may be completely defoliated during severe infestations.

 CONTROLS Use sticky tapes around tree trunks. Spray with *Bacillus thuringiensis berliner* strains.

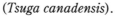

HEMLOCK WOOLLY ADELGID

Since its discovery in Connecticut in 1985, the hemlock woolly adelgid (*Adelges tsugae*) has spread throughout several northeastern states. At this time it is known to feed on only one host plant, the eastern hemlock (*Tsuga canadensis*).

Tiny and aphid-like, these sap-sucking insects usually feed on the youngest branches of a hemlock, where the needles attach to the twig. Researchers believe that adelgids inject a toxic saliva into the plant as they feed. This feeding eventually kills existing needles and interferes with the hemlock's ability to produce new ones.

LIFE CYCLE The hemlock woolly adelgid gets its name from the fact that most of its life is spent enveloped in a white woolly eggsac the size of a Q-tip. Overwintering adult females lay from 50 to 300 eggs in a single eggsac from February through June. The first eggs begin to hatch in April, and hatching continues through June. For up to 2 days newly hatched crawlers search for a new place to settle and feed. Some develop into wingless adult females that stay on the hemlock and produce another generation. Others become winged adults that fly off the hemlock to find another host plant. Those that remain lay more eggs, which hatch from June through mid-July. These second-generation adelgids remain dormant until October, when they resume development through the winter. They mature by the following February, when the life cycle begins again.

 SYMPTOMS It's easy to recognize an infestation because the eggsacs are present at the base of almost every needle. Needles discolor, dry out and eventually fall off. Often, infested limbs die within the first summer, and entire trees succumb within 1 to 4 years.

 CONTROLS The key to control of the hemlock woolly adelgid is to be very observant and not let a beginning infestation go unnoticed. Infested hemlocks must be completely drenched with any one of a number of registered pesticides using hydraulic spray equipment. One thorough application a year anytime during the entire growing season should be sufficient unless the infested hemlock is very large or dense, or other infested hemlocks are nearby. Both horticultural oil and insecticidal soap give good control of this pest.

JAPANESE BEETLE

The Japanese beetle (*Popilla japonica*) is one of the most destructive pests of lawns and gardens east of the Mississippi. The 1/2", oval-shaped beetle is metallic green with copper-colored wings. The larva, or grub, is grayish white and 1/2" to 1" long.

First observed in the U.S. in 1916 in New Jersey, the Japanese beetle has spread from Ontario and Nova Scotia, south to Georgia and west to Tennessee and Missouri. Local outbreaks have occurred in California.

LIFE CYCLE Grubs overwinter in the soil, burrowing below the frost line.

In spring, they rise to the root zone of grasses to feed. They emerge as winged adults in May and June. After 30 to 45 days, females deposit their eggs in the soil. Grubs hatch in about 2 weeks, eating grass roots until cold weather sets in. There is a single generation per year.

SYMPTOMS Adults feed on more than 200 different species of plants, rapidly skeletonzing foliage and disfiguring flowers. The larvae devour grass roots, turning large patches of turf brown.

CONTROLS Beetle traps often do more harm than good by attracting large numbers of beetles to your yard, where they are likely to prefer feeding on your plants.

Protect your plants from adult beetles by handpicking them. Like white grubs, the larvae can be controlled with commercially available parasitic nematodes. Milky spore, a bacteria that causes a lethal disease specific to the Japanese beetle grub, provides long-term control, especially when applied over a large area. Some botanical pesticide products have been registered by the Environmental Protection Agency for use against adult beetles and grubs. Check the product labels for information on which pests they control and how to use them safely.

JUNE BEETLE

June beetles (*Phyllophaga* species) are also known as May beetles, June bugs and daw bugs; larvae are called white grubs. Adults are brown, black or green beetles, about 1" long. The white grubs have a dark head, are 1/2" to 1-1/2" long and are usually curled into a "C" shape in the soil.

June beetles occur throughout the U.S., and are most serious in the South and Midwest.

Adults appear in May, June or July and feed on the foliage of hardwood trees and deciduous shrubs. Most fly at night and hide during the day in debris or dense foliage. Females enter the soil to lay eggs, which hatch in 2 or 3 weeks. Larvae feed on roots, especially lawn grasses, until cold weather forces them deeper into the soil. In the spring, they resume feeding. The larval stage lasts between 1 and 3 seasons, depending on species and location, then the larvae pupate and emerge as adults.

Adults feed at night on a wide range of plants, leaving irregular holes in the leaves.

Grub damage shows up as brown patches in the lawn. To determine if white grubs are the cause, peel back a square foot of sod in several different areas of the lawn.

Fall plowing exposes many grubs to hungry birds and freezing temperatures; moles and skunks are natural predators. Adult June beetles can be handpicked.

If you find more than five grubs per square foot, you may want to consider a more aggressive control. Commercially available predatory nematodes are effective. Some botanical pesticide products have been registered by the Environmental Protection Agency for use against the June beetle. Check the product labels for information on which pests they control and how to use them safely.

LACE BUG

Lace bugs are sap-sucking insects that seriously injure plants when they are present in large numbers. They infest a wide variety of ornamental shrubs and shade trees; they are especially damaging to azaleas, rhododendrons and andromedas (*Pieris* species).

Lace bugs are a problem from Canada to Florida, west to Ohio and Michigan, and also in the Pacific Northwest. They're active from spring through fall.

Lace bugs derive their name from the lacy appearance of the adults' dorsal surface, including the head, body and wings. Adults are about 1/8" long, and rectangular, with delicately netted, transparent wings. The spiny nymphs are nearly colorless at first, turning dark as they mature.

LIFE CYCLE Lace bugs survive the winter as eggs, generally on the underside of lower leaves. Nymphs hatch in early spring and begin feeding on their hosts. They proceed through a series of five molts in as little as 2 to 3 weeks to become adults. Summer eggs are laid on the underside of foliage near the top of plants. There are two to five generations per year.

SYMPTOMS Adults and nymphs feed in clusters on the underside of leaves. Skins shed by molting nymphs often remain on the leaves. A more obvious sign of lace bug infestation is the presence of brown, sticky spots and black excrement. The upper surface of the foliage appears stippled or blanched. Left untreated, a plant becomes unsightly, loses vigor and may die after a few seasons.

CONTROLS Insecticidal soaps applied directly onto the pests on the underside of infested leaves will help hold lace bug populations in check. Some botanical pesticide products have been registered by the Environmental Protection Agency for use against lace bugs. Check the product labels for information on which pests they control and how to use them safely.

LEAFHOPPERS

Potato leafhoppers are about 1/8" long and greenish. Beet leafhoppers are about 1/6" long and greenish, yellow or brown, usually with darker blotches. Nymphs are similar to the adults in shape, but smaller and usually a lighter color. There are many other kinds of leafhoppers, each associated with varying plant hosts, including fruit trees, grasses, ornamentals and many field and garden crops, but these are the most troublesome. Potato leafhoppers are found in the eastern half of North America. Beet leafhoppers are found in the hot, dry areas of the West.

 There are up to 4 generations per year, depending on the warmth of the climate. Potato leafhoppers migrate north from the Gulf Coast area each spring, blown by winds, and can reach southern Canada by early summer. Beet leafhoppers migrate from the warm areas of the Southwest each spring. Both are killed by fall frosts, and will overwinter only where temperatures remain above freezing. Eggs are inserted into new plant tissue. The eggs hatch into nymphs that feed and develop for several weeks before maturing into the adult stage, which feed and survive for about a month.

 The potato leafhopper feeds on beans, potatoes and alfalfa, inserting its sucking mouthparts and injecting saliva. This causes both mechanical and chemical damage to plant tissue, blocking the flow of plant liquids. On beans and potatoes, symptoms start out as a yellowing of the edges of leaves, which then curl and turn darker green. Plants become stunted and non-productive.

Beet leafhoppers feed on sugar beets, beets, tomatoes, spinach, squash, melons and many ornamental plants. Adults and nymphs of the beet leafhopper can transmit curly top virus disease while feeding, which can stunt, deform and kill plants.

CONTROLS Grow resistant varieties; potato leaf hoppers, for example, avoid beans with fuzzy leaves. Cover plants with floating row covers.

Grow pollen and nectar plants to attract big-eyed bugs, beneficial insects found throughout North America that feed on leafhoppers.

Spray insects with an insecticidal soap (the liquid must touch the insects to be effective).

LEAFMINERS

The assorted 700 North American leafminer species are the larval offspring of various small flies, beetles and moths. Most larvae look like small, 1/4"-long white grubs. The tiny 1/4" spinach leafminer is a particular nuisance because heavily infested leaves are unsightly and inedible.

LIFE CYCLE The slender gray parent flies lay white eggs on leaf undersides. Within 4 days, pale green maggots emerge and begin eating threadlike, winding tunnels within the leaves.

As feeding continues, the tunnels join together to make large, light-colored blotches filled with dark waste matter. Maggots can move from leaf to leaf. They feed for up to 3 weeks, then drop from the plant and spin cocoons in soil. Adult flies emerge in 3 weeks. Four life cycles are possible in a single season.

SYMPTOMS Circular, finger-shaped, oval or snakelike winding trails on vegetable leaves are evidence of leafminers tunneling in the incredibly small space between leaf top and bottom. Leafminers don't do a lot of harm to well established plants, although they can lower vegetable quality because they interfere with plant nutrition. Seedlings are more vulnerable, and may be killed.

CONTROLS Prevention is key; once the miner is inside the leaf, nothing works. Deep tilling helps destroy maggots. Remove affected leaves. Keep the garden free of leaf litter and weeds to prevent reinfestation.

MEALYBUGS

Mealybugs are small, white, wingless insects about 1/8" in length. The term "mealy" originates from the white waxy covering on the insect's exoskeleton. Female mealybugs use their beak-like stylets to suck sap from such plants as citrus, orchids, cacti, African violets, palms, *Ficus*, chrysanthemums, ferns and *Lantana* .

LIFE CYCLE Mealybugs can be found on leaves, stems, roots and flowers and are most common when conditions are warm and humid. Male mealybugs are small, winged insects that lack functional mouthparts. Adult females are wingless and much larger. Under optimal conditions, mealybugs can mature from the egg stage within 25 days. The eggs are often found on the underside of leaves, surrounded with white, waxy fibers. During cool conditions nymphs can hibernate in the soil and re-emerge when conditions are more favorable.

SYMPTOMS Mealybugs, like other sap feeders, exude sugary honeydew while feeding. Sooty molds often grow on honeydew deposits. These molds can significantly reduce a plant's photosynthetic efficiency and produce a less aesthetically pleasing specimen. Heavy mealybug infestation causes yellowing leaves, stunted growth and reduced flower and fruit size.

CONTROLS For single-plant infestations, remove and kill mealybugs and waxy egg sacs with cotton swabs moistened with 70 percent rubbing alcohol. Lady beetles, particularly *Cryptolaemus montrouzieri*, are an excellent biological control for mealybugs.

MEXICAN BEAN BEETLE

The adult Mexican bean beetle (*Epilachna varivestis*) is 1/4" long and orange, with 16 black dots. (Don't confuse it with the beneficial lady beetle, which has varying numbers of spots.) Bright yellow eggs are laid in clusters on the undersides of leaves. Larvae are smaller than the adults, yellowish orange, beetle shaped and covered with branched spines.

Mexican bean beetles occur in the eastern U.S.

LIFE CYCLE Adults hibernate in undisturbed woody areas or in garden debris. In spring, they fly to young bean plants to lay eggs, which hatch in 1 to 2 weeks. When the eggs hatch, the beetle larvae feed on bean leaves. The larvae form pupae on the bean leaves, and the adults emerge and feed on foliage to begin the cycle again. Depending on the warmth of the climate, there are up to three generations per year.

SYMPTOMS Mexican bean beetles skeletonize bean leaves. Severe infestations can defoliate entire plants and discolor and disfigure bean pods.

CONTROLS Keep your garden clean and cultivate it early to disturb overwintering sites. Keep bean plants under floating row covers until they are big and sturdy. Plant early-maturing bean varieties — they'll be finished before the larger populations of bean beetles appear late in the season.

Release lady beetles to feed on the beetle eggs and larvae. Spined soldier bugs, which also can be purchased, feed on the larvae, too. Grow herbs and flowers to provide pollen and nectar to attract these beneficial organisms.

NEMATODES

These microscopic wormlike animals feed on the roots, leaves and stems of more than 2,000 vegetables, fruits and ornamental plants. The best-known is the root-knot nematode, whose feeding causes the characteristic galls on roots. Other root-feeding nematodes are the cyst- and lesion-types, which are more host specific.

 LIFE CYCLE Nematodes are present throughout the U.S., but are mostly a problem in warm, humid areas of the South and West and in sandy soils. Nematodes overwinter as eggs in the soil and hatch in spring; the immature nematodes feed on nearby roots. Within a month they have developed into adults, which lay more eggs. In warm, moist locales, nematodes can be active year-round.

 SYMPTOMS The signs of nematode damage include stunting and yellowing of the leaves, and wilting of the plants during hot periods. Though not very mobile, nematodes are easily spread by water, garden tools and soil on gardeners' boots.

CONTROLS Remove infested plants, roots and all, as soon as you harvest. Send soil samples to your local Cooperative Extension office for nematode identification. Once you've identified your nematodes, you can incorporate a variety of control techniques. Growing cover crops of French marigolds, such as 'Nemagold', for 90 to 100 days then tilling them into the soil in fall, has been shown to reduce nematode numbers. In warm, sunny areas, soil solarization has been used effectively. (See the chapter on cultural and physical controls for instructions on how to solarize your soil.) Organic fertilizers and soil amendments containing seaweed and humic acids have been successfully used to reduce the numbers of nematodes in the soil while keeping the plants in optimum health.

For long-term control add organic matter to your soil to increase the number of organisms that prey on or parasitize nematodes. Grow nematode-resistant varieties (look for the "N" after the variety name in seed catalogs) and plants nematodes don't attack, such as asparagus and junipers.

NORTHERN CORN ROOT WORM

The northern corn root worm (*Diabrotica barberi*) is closely related to the cucumber beetle. It is a pest primarily in southern Canada and the American corn belt states.

LIFE CYCLE In late summer or early fall the 1/4"-long, pale green to yellowish green adult beetle lays up to 300 yellow-orange eggs in cornfield soils. The eggs overwinter and by late spring white, 1/2"-long larvae emerge and begin to feed on nearby corn roots. The northern corn root worm larvae feed specifically on corn roots, and if no roots are available, the larvae die. There is a single generation a year.

SYMPTOMS The larvae's feeding can cause corn roots to weaken and the plants to get blown over, especially during high winds. The adult beetle can feed on corn silks, causing improper pollination of the ears. The northern corn root worm also can transmit the chlorotic mottle virus, which is lethal to corn plants.

CONTROLS The simplest control is to rotate crops, planting a legume for at least 2 years in the infested soil; because the root worm attacks only corn plants, the larvae and eggs will eventually die.

Beneficial nematodes applied on the soil when eggs hatch in spring has also proven effective in controlling the root-worm larvae. These are available from mail-order suppliers.

ONION MAGGOT

This native of Europe is the most serious pest on onions and other alliums. It's mainly a problem in the northern midwestern U.S. and southern Canada.

LIFE CYCLE The onion maggot has a life cycle similar to that of the cabbage maggot, to which it is closely related.

The brownish gray, housefly-size adult emerges in April to May, depending on the area, and lays clusters of white eggs in the onion leaf axis and in the soil near the plants. When the eggs hatch, the 1/3"-long, white, tapered larvae burrow into the onion bulb.

Onions are usually planted close together, and therefore maggots can spread to other plants before pupating in the soil and emerging as adults a few weeks later. There can be 2 to 3 generations a year, but the first generation usually causes the most damage.

 SYMPTOMS By burrowing into the onion bulb, the maggots open it to infection from diseases such as bacterial soft rot, and eventually cause the plant to wilt and die.

 CONTROLS Because the adult overwinters on decomposing onion debris in the soil, proper clean-up in fall will help reduce the population the following year.

Floating row covers placed over the plants or tar paper placed on the soil around the onion bulbs in spring during transplanting prevent the adult fly from laying the eggs on or near the onions. Reflective mulch such as aluminum foil placed around the base of plants has been shown to confuse the adult flies, preventing them from finding the onions on which to lay their eggs.

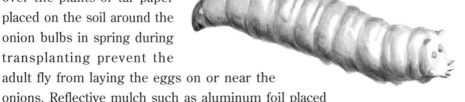

Diatomaceous earth granules sprinkled around the base of onions will kill young larvae. Repeated weekly applications may be required.

PEACH TREE BORER

Although it does not directly attack fruit, the peach tree borer *(Synanthedon exitiosa)* is the most damaging pest of *Prunus* species (peach, cherry, plum and so on) in much of North America.

The larva is a pale, grub-like insect with a distinct dark-brown head. The adults are clear-winged moths that resemble wasps.

LIFE CYCLE Partially grown caterpillars spend the winter either under the bark at the base of the tree or in small silken coverings. With the return of warm temperatures, they begin to feed on the bark and cambium of the trunk and larger roots. Peak feeding and injury occur as the developing insects mature in May and June. Pupation usually occurs just below the soil surface, in a case made of chewed wood fragments, tree gum and soil. The adult moths emerge beginning in late June and may be present through the summer. Females lay eggs in bark cracks.

SYMPTOMS Larvae of the peach tree borer do most damage at or below the soil surface. (This differs from the damage caused by the lesser peach tree borer, *Synanthedon pictipes*, which frequently attacks some of the larger, aboveground branches.) Feeding wounds girdle and weaken or kill plants.

CONTROLS You can inhibit egg laying by banding the base of trees with burlap or heavy paper, covered with a sticky paste such as Tanglefoot or Stick 'em Special. This barrier should be in place prior to egg laying and maintained throughout the summer.

Egg laying and larval attacks are often concentrated where trees have been wounded. Be careful not to wound trees with tools or other equipment.

Individual borer larvae can be destroyed with a knife or wire. This requires carefully excavating the area around the base of the trunk to a depth of several inches.

Several parasites attack developing peach tree borers and may provide sufficient control in some locations. These natural enemies are all native and can be attracted to your garden if you grow herbs and flowering plants.

Drenches of commercially available parasitic nematodes around the bases of trees have been able to control developing larvae in ornamental cherry. However, the gumming produced by larval wounding can interfere with the effectiveness of this treatment.

PEAR PSYLLA

The pear psylla (*Cacopsylla pyricola*), an insect native to Europe, was first discovered in Connecticut in 1832 and has since spread throughout pear-growing regions of the East and West.

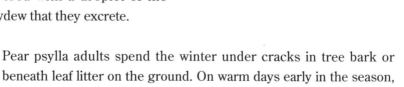

The adults are small, about the size of aphids. They look like tiny, dark cicadas and jump when disturbed. The nymphs, which do most of the feeding, are flat, greenish-brown and somewhat scale-like. They're usually covered with a droplet of the sticky honeydew that they excrete.

LIFE CYCLE Pear psylla adults spend the winter under cracks in tree bark or beneath leaf litter on the ground. On warm days early in the season, before buds break, the adults move to the twigs and branches of the pear tree and mate. Eggs are laid on the buds and spurs of the tree and, following bud break, on young leaves. Developing nymphs suck the sap from leaves, producing characteristic honeydew. Three or more generations may be produced during the course of a growing season.

Pear psylla feeds on the sap of pear leaves, reducing the vigor of the plant. On some varieties the effects of feeding are particularly severe, producing "psylla shock," which causes rapid decline and sometimes even death of affected trees. Pear psylla also can transmit the pathogen that produces pear decline disease. In addition, they excrete large amounts of sticky honeydew, which can cover fruit and leaves and on which a sooty fungus develops, detracting from the appearance and value of fruit.

 Pear psylla thrives on rapidly developing new growth. During mid-summer "water sprouts," suckers from the stem, are a major source of new growth. If you remove these suckers, you will deprive the insects of food, reducing numbers the subsequent season.

There are numerous natural enemies of pear psylla, including general predators such as green lacewings and minute pirate bugs. Several small chalcid wasps are important natural enemies of pear psylla as well. Avoid using pesticides that kill these beneficial insects.

Horticultural oils are particularly useful for controlling pear psylla when trees are sprayed early in the season, just as the buds swell.

PLUM CURCULIO

The plum curculio *(Conotrachelus nenuphar)* can damage almost all tree fruits, including peach, apricot, cherry, apple and pear as well as plum. This native American weevil is widely distributed east of the Rocky Mountains.

Larvae are plump and legless, with dark heads. About 1/3" when full grown, they're almost invariably found inside the fruit. Adults, about 1/4" long, have several light-colored patches on the back and 4 small humps. Both adult and larval stages of plum curculio damage fruit.

 Plum curculio adults spend the winter under protective debris in and around infested trees. They become active and return to trees after spells of warm weather in spring, often around the time that apple trees bloom. The adults feed on the developing fruit. Females insert eggs under the skin, then make a slit in the fruit to prevent the expanding tissues from crushing the eggs. After the eggs hatch, the larvae feed on the flesh and developing seeds of the fruit. When full grown they cut their way out of the fruit and burrow 1" to 2" into the soil to pupate.

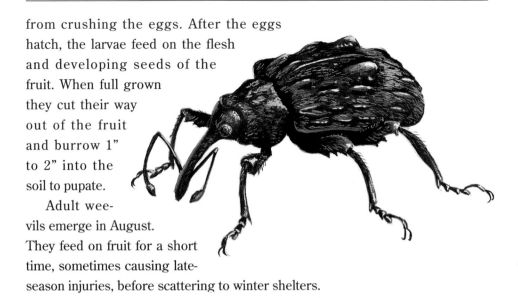

Adult weevils emerge in August. They feed on fruit for a short time, sometimes causing late-season injuries, before scattering to winter shelters.

SYMPTOMS Crescent-shaped feeding scars on the surface of the fruit are signs of injury by the adults. Feeding wounds on apples that are made early in the season often develop into sunken corky areas while late-summer feeding produces small areas of rot under the skin. Larvae feed inside the developing fruit, and infested apples usually drop prematurely. Infested stone fruit, such as cherries and plums, do not drop.

CONTROLS Shake branches to dislodge adults, collecting them on sheets spread under the tree. Because new weevils move to fruit trees over a period of several weeks in spring, this needs to be repeated several times to be effective.

Fruit that drops prematurely or that is obviously infested should be collected and destroyed to prevent larvae from completing their life cycle. This can reduce problems the following season.

Natural enemies of the plum curculio include a braconid wasp, an ichneumonid wasp and the fungus *Metarhizium anisopline*. Flowering plants in the vicinity of the fruit trees provide food sources for beneficial insects.

Some botanical pesticide products have been registered by the Environmental Protection Agency for use against the plum curculio. Check the product labels for information on which pests they control and how to use them safely.

ROSE CHAFER

The adult rose chafer (*Macrodactylus subspinosus*) is a slender tan beetle, about 1/2" long with spiny legs. The larva is a white, 3/4" grub with a brown head. It causes significant damage to roses, peonies, hollyhocks, grapes, blackberries and many other species of ornamental plants and fruit crops.

The rose chafer's preference for light, sandy soil helps limit its distribution. Most serious in the northeastern U.S. and eastern Canada, it occurs as far west as Colorado and Texas.

LIFE CYCLE Rose chafers overwinter as larvae in the soil. In spring the larvae pupate and emerge as adults in late May or June. They often appear suddenly, in large swarms, feed for 4 to 6 weeks before females lay their eggs in sandy soil, then disappear as suddenly as they came. Eggs hatch in 1 to 2 weeks; the young grubs feed on the roots of grasses and weeds until cold temperatures force them deeper into the soil. There is a single generation per year.

SYMPTOMS Adult rose chafers feed on both blossoms and leaves. Flower petals and buds are eaten and soiled with excrement; leaves are skeletonized.

CONTROLS Handpicking adult beetles is an effective control, and is relatively easy because their movements are sluggish.

Grubs can be controlled with commercially available parasitic nematodes. Some botanical pesticide products have been registered by the Environmental Protection Agency for use against rose chafers. Check the product labels for information on which pests they control and how to use them safely.

SCALES

Scales are one of the most specialized groups of sap feeders. Adult females lack legs and antennae and are usually covered with a waxy shell. A mature female in the armored scale family lives firmly attached to a twig or leaf in a waxy shell that completely encloses her body but is not attached to it. The waxy covering of the soft scale does not cover the entire body but is attached to it. The scales' body armor makes these insects difficult to control with many commonly used chemical sprays.

 Young either hatch from eggs or are born live and sheltered beneath the parent's shell. The first nymphs are crawlers and thus able to disperse to new plants or other areas on the same plant. After molting, females secrete the shell, begin to feed and remain immobile for the rest of their lives. Male development features an inactive or pupa-like stage, followed by emergence as a winged, nonfeeding adult. Male scales are very small and gnat-like.

 Scale insects are easy to spot on many plant parts, although they sometimes elude detection when they've anchored themselves beneath stem sheaths or on the undersides of leaves. Yellow spotting, decrease in plant vigor and the presence of sooty molds are signs of a heavy infestation.

 Remove scales gently with a soft-bristled toothbrush dipped in a solution of rubbing alcohol, soapy water or a combination of both.

Most scale predators prey only on specific types of scales, so proper identification is important. Your Cooperative Extension agent can help with identification. *Lindorus lophanthae*, a small beetle, preys on several species of soft and armored scales. Several small wasp species parasitize certain scale species. These are available from mail-order suppliers.

Horticultural oils are very effective at controlling scale.

SNAILS AND SLUGS

Snails and slugs occur throughout the U.S. The brown garden snail is a pest mainly in West Coast states and in parts of the South. Snails and slugs devour seedlings and soft-tissued parts of vegetables, fruits and ornamental plants.

Snails have a soft, unsegmented body in a spiral shell 1/2" to 1-1/2" across. Slugs are 1/4" to 8" long, with no shell. Young snails and slugs look much like the adults. Both lay masses of clear to white, oval to round eggs.

LIFE CYCLE Snails and slugs feed at night and on cloudy, damp days. Snails spend the day on smooth, shaded surfaces; slugs in moist crevices. Snails and many slugs are hermaphroditic, which means that every individual lays eggs. Both snails and slugs breed throughout the warmer part of the year, laying eggs that hatch in several weeks. In some cases, only eggs may overwinter, but in most cases, all life stages survive.

SYMPTOMS Both snails and slugs leave shiny slime trails. Slugs eat giant holes in favored plants, sometimes defoliating them. Snails create irregular holes in the middle or at the edges of leaves.

CONTROLS Identify favorite hiding places such as piles of pots, beneath black plastic mulch and in shaded locations on fences, walls or plants with broad, flat leaves. Remove hiding places, or inspect them regularly. Handpick daily, then weekly when numbers drop.

Homemade traps, such as boards to which narrow strips of wood have been attached, are effective; nail two 1" x 1" runners to a 1'-square board and set it on the ground with the runners on the bottom. Snails and sometimes slugs will congregate underneath. Beer or malted-grain traps also may attract slugs. Traps should be emptied daily.

Use copper screening (4" above ground, 6" to 8" underground) or copper strips, set around tree trunks or raised beds to deter these pests.

Domestic ducks (penned to protect crops), rove beetles, salamanders and moles will eat snails.

Diatomaceous earth with patented baits will attract and kill slugs without endangering pets or children.

SOWBUGS AND PILLBUGS

Pillbugs and sowbugs live in gardens in all parts of the continental U.S. They feed primarily on decaying organic matter, but if populations are large, they will eat seedlings, new roots, lower leaves and fruits lying on or near a damp soil surface. They are often blamed for damage caused by other pests, such as slugs and snails.

These crab and shrimp relatives are about 1/2" long, with segmented bodies and 7 pairs of legs. Usually dark gray, they may be lavender just after molting. Pillbugs are often found curled up into a ball; sowbugs (a different genus) are not. The young in both cases resemble the adults.

LIFE CYCLE Because they breathe through gills, pillbugs and sowbugs require a moist environment. They feed mainly at night, spending the day in dark, moist places. They mate in the spring. Females hold eggs, then young, in a pouch for more than 3 months. The young mature in 1 year, and live up to 3 years.

SYMPTOMS Sowbugs eat the roots of seedlings and girdle their stems. They chew jagged-edged holes in tender leaves and eat away the surface of thin-skinned fruits such as young zucchini.

CONTROLS Avoid overhead watering. Water early in the day so that plants and soil dry out before evening. Choose mulches, such as straw or wood chips, that let water pass easily so the soil surface can dry.

Eliminate piles of plant debris, pots, boards and the like, which harbor these insects. If you clear a new garden space, wait 2 weeks before you plant.

Traps baited with diatomaceous earth attract and kill pillbugs and sowbugs.

SPIDER MITES

Spider mites vary in size from 1/64" to 1/32" in length and are difficult to detect without a hand lens. Adults have four pairs of legs. Two-spotted spider mites are light green, and actively feeding adults have two distinct dark markings on the body. Spider mites puncture plant tissue and feed on the sap. Outdoors, spider mites infest a variety of ornamental plants. Inside the home, they prefer impatiens, orchids, geraniums and palms.

LIFE CYCLE Eggs deposited on the undersides of leaves hatch within 4-5 days. The mites develop through four stages: larva, protonymph, deutonymph and adult. Maturity is reached in 3 weeks. During unfavorable conditions mature mites can go into hibernation. Mature females overwinter in the soil.

SYMPTOMS Unfortunately, spider mites are often discovered well after populations have become established and serious plant damage has occurred. After the onset of a mite outbreak, leaves may take on a yellowish or silvery cast due to the extensive array of puncture feeding scars on the undersides of leaves. Severe yellowing and rusty spots on upper leaf surfaces, followed by premature leaf death, occur in most cases of spider mite infestation. The mites also produce fine silk webbing that can cover leaves and stems.

CONTROLS Spider mites and their webbing can be washed off with a mild detergent and a water rinse. Pruning heavily infested branches can slow the spread of these pests, as can isolating infested plants.

Many species of predacious mites, including *Phytoseiulus persimilis, Mesoseiulus longipes, Neoseiulus californicus* and *Galendromus occidentalis,* are commercially available for control of two-spotted spider mites. A combination of two or more predator species provides optimum control.

Horticultural oils and insecticidal soaps provide some control.

SQUASH BUG

The squash bug (*Anasa tristis*) is present throughout the U.S. Pumpkin and squash are its primary victims, and to a lesser extent melons and gourds. The bugs suck out plant juices, and at the same time inject a harmful substance that causes rapid wilting.

The 5/8"-long, flat, hard-shelled adults are brownish black. Abdominal edges, seen under the wings, are yellow-orange.

LIFE CYCLE Adults overwinter in crop debris, or even on nearby boards. They fly to host plants in late spring. Mating and egg laying occur shortly afterward.

Each female places about 100 to 300 orangish eggs on the undersides of the host plant's leaves. Hatching occurs in 7 to 14 days. At first young are pale green with pinkish head and legs. They become grayish with black legs. The adult form occurs at 6 weeks. There is a single generation per year.

All stages can be found on crops, sometimes as many as 100 on a single vine. When numerous, squash bugs cause a disagreeable odor around the plants. This comes from special glands on the insect's abdomen. If you squeeze or crush the bug, the odor will come off on your hand.

 SYMPTOMS An infested squash plant wilts from the intial point of attack, where brown specks are likely to form, to the end of the vine. Affected vines become black and crisp. No squashes may form. Smaller plants may die. Look for squash bugs on the undersides of leaves and fruits.

 CONTROLS Adult squash bugs hibernate over the winter under garden debris. Remove and destroy unused fruit and old vines.

Try placing a piece of board between rows of the host plants. Adults and juvenile squash bugs seek shelter under these at night. In the early morning, turn the boards over and kill the sleeping pests.

A species of tachinid fly is up to 90 percent effective against squash bugs in the eastern U.S. These beneficial flies feed on herb flowers, so plant herbs in and around your garden.

SQUASH VINE BORER

The squash vine borer (*Melittia cucurbitae*) is a nuisance only east of the Rocky Mountains. Squash and pumpkins are the plants most often damaged by this pest.

LIFE CYCLE The borers overwinter in tough, 3/4", black cocoons in soil near vine crops. When host plants begin producing runners, colorful 1" borer moths emerge. They have shiny copper-green front wings, transparent hind wings and an orange and black abdomen. However, they don't really look like moths because their back wings are transparent.

These wasp-shaped moths fly swiftly in a zig-zag pattern around plants during the day. Each female eventually glues 200 pin-head-size brown eggs to plant stems, leaf stalks and fruits.

One week later, larvae emerge, 1"-long, wrinkled, fat white caterpillars with a brown head. There can be over 100 caterpillars in a single vine. The larvae begin boring into plant stems and fruit. In large numbers, they can destroy entire plants. After 6 weeks, they come out of the stems and burrow about 2" deep into nearby ground. Here they enter the cocoon stage for another winter season. They become brown pupae in early spring and then emerge shortly afterward as adult moths.

There is usually a single generation per year.

SYMPTOMS The first sign of borer is sudden wilting of squash vine runners in midsummer. Soon, yellow, sawdust-like waste matter protrudes from stem holes and appears on nearby ground. In sufficient numbers, the borers can cut off the plant's supply of food and water and kill it.

CONTROLS Prevention is the key, because once borers get inside the stem they are difficult to control. Cultivate the soil deeply. Grow resistant squash varieties. The borers can overwinter in the soil, so move your squash beds every year. Plant early, because mature plants fend off damage better than young ones do. Use row covers to prevent infestation. Remove all dead vines quickly.

TARNISHED PLANT BUG

The adult tarnished plant bug *(Lygus lineolaris)* is about 1/4" long. Its color ranges from light brown to black, with a triangle pattern on the thorax. The nymphs are 1/32" long and light yellow to green. Eggs are white.

The tarnished plant bug is found throughout North America, especially on the West Coast. It feeds on hundreds of plants, including flowers, fruits and vegetables. The bug is usually attracted to the tender shoots and growth points of plants and often feeds on flower buds as well, especially blooms of fruit such as apples, peaches and strawberries.

LIFE CYCLE The adult tarnished plant bug overwinters in garden debris, under stones, boards or other potential shelter. It emerges in early spring, mates and begins feeding. Emergence is timed so that many flowers are in bud or about to bloom just as the bugs are ready to feed.

Within about a month, the females insert eggs directly into plants, which itself causes damage to the plant. The nymphs emerge in 7 to 10 days and begin feeding. The light-green immature bugs look like aphids but they are more active. They will be fully grown within a month and lay eggs. There are up to 3 generations per year.

SYMPTOMS Both nymphs and adults injure crops by sucking out the juices. On many plants, a small black spot appears where the insect has been feeding, which causes a deformation of the stem or leaf.

Tarnished plant bug injury shows up in subsequent fruit. Strawberries, for example, will have hard, seedy ends; apples may be cat-faced, small and shrunken. Raspberries that form fruit from buds affected by the bug may be crumbly and small. On ornamentals, flowers may be small or poorly developed.

CONTROLS Cultural controls can help to lessen the severity of tarnished plant bug infestations. Clean up the garden as much as possible in fall to eliminate overwintering sites. Keep the garden as weed-free as possible, as the bugs congregate in pigweed and lambsquarters.

Cover row crops such as strawberries with a floating row cover before bloom to keep the pest away. Apply the row cover as early as possible in the season and fasten the ends securely to keep the pest from creeping under. Keep in mind that insect-pollinated plants must be uncovered when the plant blooms.

Strawberry farmers in the West have had great success with vacuuming the pests from fields using huge, tractor-pulled vacuums. Gardeners can approximate this measure with hand held, battery-powered vacuums.

A braconid wasp species parasitizes the tarnished plant bug. Grow herbs and flowers in your garden to attract this and other beneficial insects.

Some botanical pesticide products have been registered by the Environmental Protection Agency for use against the bug. Check the product labels for information on which pests they control and how to use them safely.

TENT CATERPILLARS

The tent caterpillars are common and destructive pests of cherry, apple and other tree members of the rose family. The eastern tent caterpillar (*Malacosoma americanum*) and forest tent caterpillar (*M. disstria*) range throughout southern Canada and the U.S. east of the Rockies. The western tent caterpillar (*M. californicum pluviale*) is found from the Rockies westward.

LIFE CYCLE Eggs are laid in midsummer in large, dark, glossy, spindle- or barrel-shaped masses encircling the twigs of the host plant. The eggs overwinter, hatching into larvae in early April in the Northeast. Mature larval eastern tent caterpillars are bristly, black with yellow stripes and dotted with blue along the sides and white along the back. Forest tent caterpillar larvae are bristly, blue with black and orange stripes and flecked along the back with keyhole-shaped white spots.

Both species pupate in white or pale-yellow cocoons spun on trees, fences, walls or in folded leaves. The annual cycle is completed in late June and early July, when the adults, inconspicuous dusty-brown moths, emerge to mate.

The larvae of both species are gregarious, forming silken nests from which they venture forth to feed on dry, sunny days. Nests of the eastern tent caterpillar are spun in the forks of branches; those of the forest tent caterpillar, as flat mats on trunks and branches. Nests increase in size as the caterpillars grow and are conspicuous by late spring. Leaves are consumed from the edges inward. Entire branches and sometimes trees may be defoliated.

Prune away and destroy egg masses and the tents of the eastern tent caterpillar. Spray with *Bacillus thuringiensis berliner* strains.

THRIPS

Thrips are tiny insects, 1/50" to 1/25" long. They have slender, elongated bodies with 2 pairs of narrow, bristle-like wings (which they don't use to fly). When the insect is at rest the wings lie together along the back. The body is pale yellow tinged with black in the onion thrips, and brown to black in the flower thrips.

Thrips are distinctive among insects. Their mouth parts are unlike any other insect's, sharing characteristics of those of both biting and sucking insects. They feed by scraping the surface of the plant, releasing sap, which they ingest.

Flower thrips are strongly attracted to roses and peonies. They especially like light-colored flowers. They creep into the buds and begin feeding on them. The buds may brown out and refuse to open; if they do open, the flowers may have brown edges.

LIFE CYCLE Thrips overwinter as eggs. The eggs are less than 1/100" long, too small to see with the unaided eye. They are laid singly, just below the surface of a leaf. They hatch in early spring, or about 4 days after they are laid when the weather gets warmer. The young nymphs resemble adults in shape but are almost transparent at first, changing to greenish-yellow. They are frequently found feeding in small groups. Within 2 to 3 weeks the insects undergo several molts, emerging as winged adults to repeat the life cycle.

SYMPTOMS As thrips scrape and feed, they leave streaks or black spots on the plants. Thrips are a major problem on onions. They make long, thin cuts as they feed on onion leaves. The effect is a condition called "silvering" — the entire plant takes on a streaked, dull, grayish cast.

CONTROLS Eggs of thrips overwinter in host plants, so make sure to clean up onions and other crops after frost.

Thrips can be trapped with sticky yellow traps. They can be slowed by spraying plants with water frequently. In severe cases, spraying with insecticidal soap is effective. Some botanical pesticide products have been registered by the Environmental Protection Agency for use against thrips. Check the product labels for information on which pests they control and how to use them safely.

TOMATO HORNWORM

The tomato hornworm *(Manduca quinquemaculata)* is 3" to 4" long when fully grown. (And it is fully grown only 3 weeks after hatching from eggs in late spring or early summer.) Tomato hornworms are dark green in color, often perfectly camouflaged in tomato vines, with white stripes on the sides of their bodies. They get their name from the stout horn that emerges from the tip of the abdomen. The horn is black in the northern species and red in the southern.

LIFE CYCLE The worms pupate in the soil, and after about 3 weeks the adult moth emerges. The moths, with a wingspan of about 5", are brown with orange spots on the abdomen. They lay spherical yellow-green eggs on the underside of tomato plant leaves. There may be 2 generations in the South.

SYMPTOMS Larvae chew leaves and sometimes fruit. They can quickly skeletonize a tomato plant. Look for black, buckshot-sized droppings at the base of the plant.

CONTROLS Handpicking is the most common and effective control method for small plantings. Remove the larvae and destroy them. However, if you see numerous small white cocoons on the back of a larva, leave it, because that means it has been parasitized by the tiny braconid wasp. Allow the wasps to mature and they will infect other hornworms.

Small hornworms may also be controlled with *Bacillus thuringiensis*.

As a preventive measure, cultivate the soil deeply in the fall to destroy the pupae.

WEEVILS

Weevils belong to the order Coleoptera — the beetles — but possess a unique, elongated snout with jaws at the tip. Adults may vary in color, shape and size, though most are less than 1/2" long. Larvae are generally pale, legless and grub-like.

Weevils are serious pests of plants and stored grains.

Among the most destructive plant feeders are the black vine weevil, bean weevil and cotton boll weevil.

Most species produce 1 or 2 generations per year. Larvae overwinter in the soil or in stored seed. Occasionally adults overwinter under garden debris. Eggs are usually laid near the base of susceptible plants, or shallowly deposited in plant tissue. After hatching, most larvae feed on plant roots or lower stems. In the case of the bean weevil, eggs are laid in the pod; hatching larvae eat their way into the bean seed.

 Weevil larvae can kill plants by destroying feeder roots and girdling stems. The cotton boll weevil is responsible for enormous losses in the cotton industry by damaging buds and preventing flower development. Adults also cause damage. Adult black vine weevils feed on the foliage of numerous plants, especially yews and broadleaf evergreens such as rhododendrons and euonymus. Feeding at night, they cut holes all round the leaf margin or devour entire leaves, except the mid-rib and large veins.

 To control bean weevils, remove garden debris, rotate crops and dust with diatomaceous earth.

Commercially available parasitic nematodes are a useful deterrent for root-feeding larvae. Some botanical pesticide products have been registered by the Environmental Protection Agency for use against these weevils. Check the product labels for information on which pests they control and how to use them safely.

WHITEFLIES

Most species of whitefly originate in the tropics and are largely a greenhouse and houseplant pest in the northern hemisphere. Some species are known to feed on field-grown crops of vegetable, fruit and ornamental plants. Adult whiteflies are small, 1/16" flying insects that have two pairs of opaque white wings. They have sucking mouthparts and feed heavily on plant sap. Whiteflies usually rest and feed on the undersides of leaves and take flight when disturbed. They are known to attack fuchsias, hibiscus, begonias, geraniums, cucumbers, citrus, tomatoes, potatoes and coleus.

LIFE CYCLE Whitefly females lay their eggs on the undersides of leaves. The stalked eggs of the greenhouse whitefly are often deposited in a circular pattern. The hatchling nymphs are very active while seeking a feeding site. Later life stages are scale-like and continue growing for at least 30 days, causing extensive damage. During the second to the last nymph stage, known as the pupal stage, wings develop and feeding ceases. Adults emerge to feed on the host plants. They are strong fliers and move readily among plants.

SYMPTOMS Affected plants have wilted leaves and are often coated with sooty molds. Whitefly attack reduces plant vitality, causes premature leaf death and may ultimately kill the plant.

CONTROLS Rinsing plants with a water spray disrupts feeding and dislodges eggs, nymphs and pupae as well.

Because whiteflies do not tolerate cool temperatures, lowering growing temperatures may decrease whitefly activity.

Encarsia formosa is a whitefly predator that is an effective control for whitefly populations on cultivated plants. This small wasp, which works best in warm, humid conditions, is available from mail-order suppliers.

WIREWORMS

These tough-skinned, segmented, soil-dwelling larvae are 1/2" to 1-1/2" long. They have 3 pairs of short legs near the head end and are typically white or yellow. Pupae are naked, often yellowish. Adults are brown to black beetles, under 1" long.

Wireworms are found throughout the U.S., and are especially common in irrigated parts of the West. They are most active in warm, moist soil and migrate deeper when soil is hot and dry. Wireworms eat seeds, seedlings and roots of potato, bean, beet, carrot, corn, lettuce, onion, sweet potato, grasses, asters, gladioli, dahlias and other plants. Different species have different favorites.

LIFE CYCLE Adults overwinter in the soil, then lay eggs singly 1" to 6" deep in the soil. Larvae hatch in a few days to a few weeks. Some species mature in a year, but most feed for 2 to 6 years, then pupate in late summer. You are likely to see all three stages.

SYMPTOMS Wireworms eat narrow tunnels into roots and tubers. These tunnels can be a point of entry for decay.

CONTROLS Avoid planting where turf has grown in the past year. Handpick all stages if you see them. Identify the wireworm in your garden (your Cooperative Extension agent can help you with identification), research its food preferences and rotate with resistant crops. Till the soil several times from midsummer to fall; the disturbance kills some wireworms in all life stages and exposes others to birds and other predators. Bury roots of a susceptible crop, such as carrots or potatoes, to act as lures, then dig them and discard them with the worms.

Apply commercially available beneficial nematodes to the soil.

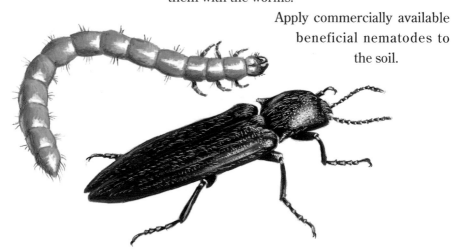

CULTURAL AND PHYSICAL CONTROLS

The first line of defense against garden pests — and the most environment-friendly

BY JACK RUTTLE

The techniques described here are the absolute first line of defense against garden pests and diseases. Many of the practices that fall under the "physical and cultural" rubric are old-fashioned, good-gardening common sense. They're things you do before you even think about mailing off an order for ladybugs or checking to see if the gaskets in the sprayer are up to another season of use. Of all pest-control methods, these are some of the most environment-friendly.

CULTURAL CONTROLS

These are an assortment of good gardening practices that should always be used, whether pests are present or not. Failure to do these things sets the stage for some serious insect and disease problems. Make the following cultural controls a part of your gardening routine.

1 CHOOSE THE RIGHT PLANT FOR THE SITE

Make sure the plants you select are suited to the soil, moisture and other conditions in your garden and receive the proper amount of sun or shade. For example, in the wild azaleas typically are shaded by the canopy of surrounding trees, so planting most azaleas in full sun instead of the dappled shade to which they are adapted is an open invitation to the lacebug. Plants that match your site conditions will be healthier than those that don't, and therefore better able to withstand depredation by pests.

Many pests spread like wildfire from one plant to another. Mixing up the crops can slow down the pest population explosion. Rows of broccoli and lettuce are alternated in this garden.

2 LOOK FOR RESISTANT VARIETIES

Plant breeders have produced a good number of plants that are resistant to diseases and, to a lesser extent, insects. To find these varieties, study catalog descriptions for references to the ability to ward off pests. For example, some vegetables have good resistance to nematodes, the microscopic worms that infest roots. Some species of birch, including paper birch and black birch, are less susceptible than others to the bronze birch borer.

3 ROTATE CROPS

It's a good idea to move most crops around the vegetable garden every year if you have the space. Many disease organisms live in the soil or on the soil surface. Likewise, several major insect pests restrict themselves to a relatively small patch of ground, traveling only short distances in their search for food. Some of these insects and diseases can overwinter in the soil. Planting the same crop in the same spot year after year encourages these pests.

Plants that are in the same families tend to be susceptible to similar insects and diseases. When planning rotations, treat all kinds of beans as essentially the same variety. Eggplant, potatoes, tomatoes and peppers are all in the nightshade family. Cabbage, broccoli, cauliflower, kale and Brussels sprouts are basically the same in the eyes of most pests. Spinach and beets are close cousins. Carrots, parsley and dill are closely related. Cucumbers, melons and squash are all prone to the same pests. Move all of these vegetables around the garden in groups.

4 MIX UP THE PLANTING

This is another prudent practice for vegetable gardeners with plenty of room. Unfortunately, there seems to be little truth to the cherished old idea that certain plants — like tansy or garlic or marigold — have a mysterious power that protects neighboring plants that we like to eat from insects. But there are other reasons for mixing plants of one kind among others. Plant diseases spread like wildfire from one plant to another. So do some insects. They move readily right down a row or through a field. But if you break up the planting with flowers or unrelated vegetables, the pest population explosion can be slowed down. Mix plants of similar sizes and shapes and season of ripening to avoid shading and make the most efficient use of space.

Beneficial flies, wasps and lacewings seek out sources of pollen and nectar for themselves, then look for nearby pests in which to lay their parasitic eggs. So it is a good idea to mix food plants for them in with your vegetables. Dill, lovage, thyme and the mints are favored food sources for many beneficials.

5 CLEAN UP THE GARDEN

Many insects and diseases overwinter on the debris of the plants they plague in summer. It's just good sense to gather up spent stalks of vegetables, flowers and weeds and cycle them through the compost heap. You should also compost the mulch you've applied under the plants in the annual flower and vegetable gardens. The ideal next step in these areas would be to plant a cover crop. Working the ground to make a seed-bed for the cover will kill some forms of wintering insects, and expose others to the elements as well as foraging birds and other predators. The cover crop itself will enrich the soil. If you can't plant a cover early enough for germination or growth, the next best thing is to apply a fresh mulch, which will prevent winter rains from compacting the soil and leave the soil ready for planting with minimal preparation when spring arrives.

6 PLANT AT THE RIGHT TIME

By planting some vegetables early or late you can avoid the worst infestations of some pests. For example, potatoes do well in cool weather, and early-planted potatoes can get strong enough before beetle populations peak to withstand a moderate infestation and still yield a respectable crop. See the Encyclopedia of Pests for additional planting tips.

7 SOLARIZE THE SOIL

You can take advantage of the sun's rays to heat the soil to a level that kills many diseases, insects and weed seeds. This is a relatively new practice that works best where the weather is sunny. The main problem for most people is that you need to be able to take a patch of ground out of production for at least a month during the prime growing season. If cloudy weather moves in, the solarizing period may be even longer. However, if you have a bad soil disease problem, or nematodes or serious infestations of weeds, you may want to give this method a try.

To solarize your soil, cover the patch of ground with a sheet of clear polyethylene. First you need to till the soil thoroughly. The plastic should be in as close contact with the soil as possible, so the surface tilth should be fine and smooth. The dark soil is your solar collector, so after you've prepared the ground, wet it; water darkens the surface and the heat travels deeper into the ground. Then put on the plastic, burying the edges in a shallow trench along the perimeter. It's good to monitor the temperature with a soil thermometer. The soil needs to get into the 150°-plus range.

When the process is complete and you begin to plant, disturb the soil as little as possible. Below the 3" to 5" depth, you will begin to encounter weed seeds that are still viable, but if they remain buried, they'll stay dormant.

PHYSICAL CONTROLS

Let's be blunt: These are practices that crush, drown, burn, incarcerate or exclude pests from garden plants. The most basic and most potent physical control agent of all is you, with your sharp eyes, powerful fingers, quick mind and big (to a bug), heavy feet. To put a prettier face on all this, it's worth quoting an old saying that the best fertilizer of all is the footsteps of the gardener. Of course, the person making those footsteps is not there fertilizing but simply attending to the plants, looking for problems and nipping them in the bud.

The most basic and potent pest control of all is you. Get out in the garden regularly to look for problems so you can nip them in the bud. More times than not you can solve the problem the minute you see it by handpicking — that is, picking off the offending insect and either crushing it or dropping it into a container of soapy water.

The first step is to identify a specific pest or disease as the culprit. Do nothing until you have identified the problem creature. If possible go into the garden once a day. When you are really stumped by damage you see on flowers or foliage, take a magnifying glass into the garden for a close look. Or sit quietly nearby for five to 10 minutes to see what moves when you stop moving. Squirt the plant with a stream of water to roust insects hiding beneath the foliage. And occasionally go out at night with a flashlight to see who is "working" your garden on the night shift.

HANDPICKING

More times than not, you can solve the pest problem the minute you see it, by picking it up and pinching it between thumb and forefinger — the yuckiest but still most popular insecticide of all.

Handpicking works best on the slowest insects, like those still inside their eggs (often in masses on the undersides of leaves) or caterpillars. Once you see cutworm damage, you can often find them by scratching around in the soil near the most recent destruction. Japanese beetles and Colorado potato beetles are also fairly sluggish. Catching cucumber beetles or flea beetles, on the other hand, can be difficult because they are extremely nimble.

Increase your chances in the handpicking game by going out in the coolest part of the day. Use a squirt-gun full of mildly soapy water to dowse the wings of flying insects to slow them down. Create places for pests to congregate, then visit daily to put an end to the party: cucumber beetles will gather under wilted squash vines, slugs will gather under a piece of board, and earwigs will gang together inside tubes of rolled-up newspaper or hollow sections of bamboo.

Gather insects in containers of soapy water, where they will drown. (When the containers are full, you can dump them on the compost pile.) Older gardening books often recommend cans of gas or kerosene. But these are petrochemicals, not-so-distant relatives of synthetic pesticides. They hold no advantage over soapy water in killing the insects, are difficult to dispose of and make a mess if spilled.

Rowcovers made of thin, lightweight polyester fabrics let in rainwater and light but are impenetrable even by small insects. Remember that the covers need to be removed from some crops for a couple of hours a day to permit pollination.

2 BARRIERS

ROWCOVERS The thin, lightweight, spunbonded polyester fabrics developed in recent years for vegetable farmers and gardeners let in rainwater and plenty of light for growth but are impenetrable even by small insects. Rowcovers were first marketed as frost protectors, but are getting more and more use over raised beds to exclude flea beetles, cucumber beetles, maggot flies and the like. The material is draped loosely over a patch, raised bed or single plant. It is so light that the plants lift it easily as they grow; there's really no need for metal hoops or other structures to hold the fabric up. Remember, though, that the covers need to be removed from some crops long enough to permit pollination.

CUTWORM COLLARS Cutworms are caterpillars that hide by day and come out at night to encircle the stems of seedlings, tender plants and even the leaves of trees. In the case of transplants, they eat away a ring of tissue, and the plant topples. However, it's easy to protect your transplants by putting paper collars around their stems.

The collars don't have to be very big. They should be about 3" high and must encircle the plant. They can be made of newspaper, tarpaper, cardboard or even plastic, which can be reused year after year. About half of the collar should be underground.

ROOT-MAGGOT SHIELDS To prevent cabbage maggot flies from laying eggs at the base of young transplants of cabbage, broccoli and other members of the mustard family, make 4" to 6" squares of tarpaper. Punch a little hole in the center, then make a slit from the hole to the edge of the square. After planting cabbages, broccoli or any other brassicas, slip the tarpaper shield over the ground at the base of the plant. The maggot can't get at the soil near the stem to lay its eggs.

TREE WRAPS A band of sticky material on the trunk of a tree will prevent some pests such as gypsy moth larvae from climbing the trunk. About a foot off the ground (that is, above the tips of surrounding foliage or flowers) tie on a heavy paper or plastic band, making it tight both top and bottom, then cover it with Tanglefoot, a thick petroleum goo, or comparable product. Don't spread the sticky stuff directly on the bark, especially on young trees.

FRUIT BAGS One of the simplest ways to prevent insect and disease damage to apples is to fasten a paper bag around each one. Admittedly, the tree is no longer quite so attractive; on the other hand, the result is extra-fancy fruit. Buy small brown paper bags, or cut up the long thin kind that wine bottles come in. The method also works well on low-lying tomatoes to prevent slug damage. For grapes, use larger bags.

3 TRAPS

STICKY YELLOW BOARDS Aphids, thrips and whiteflies are attracted to bright yellow. Paint a square of plastic or wood yellow (the ideal color is Rustoleum

Traps let you know when pests arrive in your garden. In some cases they can keep pest populations at acceptable levels. Different color traps work for different pests. Aphids are among the insects attracted to yellow traps, while red spheres attract apple maggot flies. Blue traps monitor populations of flower thrips. Japanese beetle traps, bottom left, are baited with sexual and floral attractants. Place these traps well away from the garden or the beetle infestation may get even worse.

Safety Yellow), cover it with Tanglefoot and place it near the plants under attack to permanently lure away some of the horde. Place the traps at foliage height and within a foot of the plant where the pests are most dense. Shake the plant and many pests will fly directly to the trap.

JAPANESE BEETLE TRAPS These traps are baited with both a food-scent lure and a sex pheromone lure. The beetles fly in, hit a metal baffle and fall into a bag, which can be emptied periodically. The beetles are strongly attracted to the trap, but are just as happy to land on a nearby food source and begin mating there. So place it far away from the plants you want to protect, a hundred feet or more. Empty the traps several times a week. Japanese beetle traps are sold in almost any garden center or hardware store in spring.

PHEROMONE TRAPS Sex pheromone traps are available for gypsy moths and several kinds of fruit pests. However, these traps attract only male insects and do not provide good control. They are really intended to help monitor a growing insect problem so that people can predict the ideal time for taking other control measures like spraying.

SLUG TRAPS Slugs are strongly attracted to the smell of alcohol and other fermentation byproducts, and will gladly crawl into a dish of beer or sugar water and yeast. These traps should be emptied of drowned slugs daily. Set the traps in a depression so they are almost flush with ground level. (They should be raised ever so slightly so that soil doesn't get in.) Rig up a shingle or piece of board to keep the sun and rain from exhausting the brew.

4 DIATOMACEOUS EARTH AND SPRAYS OF WATER

You may think of these as sprays or dusts because that's how they're applied. But they are physical controls, not poisons. A strong spray from a hose will knock off aphids and spider mites, and they'll be unable to crawl back onto the plant. Diatomaceous earth (D.E.) is mined from deposits of the microscopic skeletons of diatoms. It is applied dry with a duster, and kills soft-bodied insects like whitefly and aphids by physically scratching them and causing dehydration. Use a mask to avoid damage to your lungs. And be aware that some D.E. products are mixed with pyrethrum, a botanical poison.

THE PREDATOR PATROL

Beneficial insects in your garden can help keep the bad guys under control

BY JOANNA PONCAVAGE

If you've ever seen a tomato hornworm sicken and die, its back thick with little white knobs that look like grains of rice, you've seen natural pest control at its most graphic. Those little white grains are pupae (cocoons) that have developed from eggs laid on the hornworm by a braconid wasp. As the wasps develop, they kill their host. Then they go looking for more hornworms.

We call these tiny wasps "beneficial" because what they do naturally has horticultural benefits. (An added benefit is that these wasps don't sting humans.) To gardeners, beneficial insects are the "good" insects. They help us grow fruits, vegetables, trees and flowers by eating or parasitizing pest insects. Almost every insect has a specific set of natural enemies — other insect species that will eat it, or use its body or eggs as a place to lay their own eggs.

Thanks to a growing awareness of the dangers of chemical insecticides, there's been increased interest in beneficial insects and other beneficial organisms, such as mites and nematodes. Many scientists are studying how best to put insects to work in our gardens, fields and orchards. Now you can buy lures that entice beneficial insects to your garden with sexual smells (called pheromones) or food aromas. You can also buy insects such as lady beetles and lacewings, two of the most common (and voracious) beneficials.

Using beneficials doesn't get the immediate results that spraying an insecticide does. It also requires more understanding and observation of what's going on, bug-wise, in your garden. The goal of using beneficial insects is not total eradication — rather, it is to provide a natural balance between pests and their predators so that the populations of bad guys don't get out of control.

So how do you go about using beneficials in your garden? First, identify the insects that are causing problems. (A magnifying glass and a good insect identification field guide are helpful.) Not every beneficial eats or parasitizes every pest, so you need to match the correct beneficial or beneficials to your particular pests. Timing is important, too. If you want to release an egg parasite, you must do so when the pest to be controlled is in its egg stage. And if you buy hungry beneficials, they've got to have something to eat when you set them free.

The following are some of the most commonly available beneficial insects. They're also some of the least fussy eaters, preying on a variety of pests.

BENEFICIAL INSECTS FOR THE GARDEN

LADY BEETLES Better known as ladybugs, these are the most widely recognized beneficials. The species that is usually sold is the convergent lady beetle — the classic red beetle with black spots. Its favorite food is aphids, but it will help control other soft-bodied pests, too. When you buy lady beetles, you'll receive the adult stage. Open the container and sprinkle them with a little water to give them a drink, then close them back up and keep them in your refrigerator until conditions are right for their release. Don't set them loose in the middle of a hot day; wait until evening, preferably after a rain. (Water your garden first if it hasn't rained.) And make sure there is something in your garden for them to eat, preferably aphids, but pollen and nectar of blooming herbs and flowers will do. You want the adults to stay around long enough to lay eggs. These will hatch into voracious larvae that will eat up more aphids by the hundreds.

LACEWINGS The lacewing is found naturally throughout North America. The larvae are big aphid eaters, too, but they'll also feed on thrips, mealybugs, scale, moth eggs, small caterpillars and mites. If you buy lacewings, you'll receive the egg stage. Scatter about 1,000 lacewing eggs around each 500 square feet of garden.

SPINED SOLDIER BUGS These beneficial bugs are also common throughout North America, but you can buy them, too. They are shield-shaped, yellow or brown bugs about one-half inch long. After the nymphs

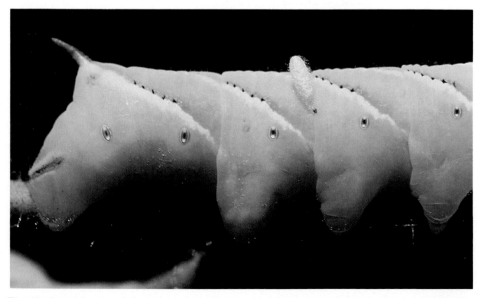

The little white knob is a sign that this tomato hornworm is in the early stages of parasitization by a braconid wasp.

hatch from their eggs,
they become voracious
predators.
They will eat
many kinds of
caterpillars, including gypsy
moth caterpillars, plus
Colorado potato
beetle larvae and Mex-
ican bean beetle larvae.
You'll need about 100
spined soldier bug
nymphs per 20 feet of
bean row, or two to five
nymphs for each square yard.

Continues on page 92

The number of knobs, technically pupae, have increased. The hornworm's days are numbered.

HOW TO ENTICE BENEFICIALS TO YOUR GARDEN

There are many beneficial insects that you won't find for sale in your favorite gardening catalog, even though they are some of the most effective. Big-eyed bugs, for example, can eat up to 12 small caterpillars or leafhoppers per day; damsel bug nymphs eat aphids, thrips and more; minute pirate bug nymphs will eat many insects, including thrips, spider mites, small caterpillars, leafhopper nymphs, corn earworms and the eggs of many insects.

In fact, you might not have to buy beneficial insects at all. Naturally occurring beneficials will do a good job of bugging pests in your garden if you make it insect-friendly. To create an enticing habitat for beneficials:

1 Grow plants that will provide nectar and pollen for beneficial insects. This will attract them and give them something to eat while they wait for the pest insects to arrive or hatch. Many common herbs such as dill, caraway, fennel, spearmint and lemon balm are beneficial favorites. Tansy (*Tanacetum vulgare*, the herb, not tansy ragwort, the weed) is especially attractive to lady beetles. Other plants favored by beneficial insects include white cosmos, clovers and other legumes, buckwheat and many wildflowers. Stagger your plantings to make sure one or more of these plants is always in bloom in your garden.

2 Provide a source of water. This can be a small pond, or a shallow dish filled with pebbles and water.

3 Give your beneficials some shelter from wind and rain with a hedge of tall sunflowers or perennials. A perennial border also provides food and overwintering sites. Cover crops add organic matter when you turn them into the soil, and will provide shelter and food for beneficials as well.

4 Lay off the pesticides. Many of them kill beneficial insects as well as pests. If you must use pesticides, use products that do not harm populations of the "good guys" in your garden.

Grow herbs, flowering plants and legumes to attract beneficial insects to your garden. Fennel (inset) and other plants in the carrot family are particular favorites. Be sure to lay off the pesticides! Many of them kill beneficial insects as well as pests. if you do use them, look for products that don't harm the "good guys" in your garden.

TRICHOGRAMMA WASPS These are tiny insects that will lay their eggs in the freshly laid eggs of many notorious garden pests, including corn earworms, cutworms, cabbage loopers, codling moths and tomato hornworms. The wasp larvae develop inside the host eggs, then eat the contents of the eggs, which turn black as the wasps mature. Trichogramma wasps do not sting humans. They are native to North America but can be purchased from farm supply companies.

NEMATODES These are extremely tiny worms that live in the soil. Some nematodes are parasitic on insects that live at least part of their lives underground, including cutworms, root weevils, wireworms and white grubs. When you buy beneficial nematodes, you'll receive a small container that holds millions of them. First, rake back any mulch or thatch from the area you intend to apply them to. Then water thoroughly until the soil is very moist — the wetter the better. Mix the nematodes with water according to package instructions. Use a watering can to sprinkle them where you want them, again noting package instructions for the rate of application. Then water the area again to soak the nematodes into the soil. This whole process is best done in the evening, because exposure to sunlight and dry air can kill nematodes.

The praying mantis is also widely available for sale as a beneficial insect. However, it has fallen into disfavor as entomologists have told us how truly omnivorous this insect is. It will eat anything — including other beneficials and its own young. In other words, it does as much harm as good.

BENEFICIAL INSECTS FOR THE HOME AND GREENHOUSE

If you have a greenhouse or lots of indoor plants, you may have problems with aphids, whiteflies, spider mites, mealybugs or scale. Greenhouses are ideal battlegrounds for beneficials — closed environments from which they can't escape. Lady beetles are very effective in greenhouses, preying on most of the above greenhouse pests.

ENCARSIA FORMOSA Tiny wasps with the scientific name of *Encarsia formosa* parasitize the pesky whitefly. The wasps lay their eggs in immature whitefly or sweet potato whitefly larvae. As the wasp larvae develop inside the whitefly, the whiteflies die. These little wasps will not survive cold winters, but thrive in greenhouses and mild-winter areas, or can be released outdoors in colder locales when the weather is warm. About 1,000 wasps should be enough for most home gardens or greenhouses.

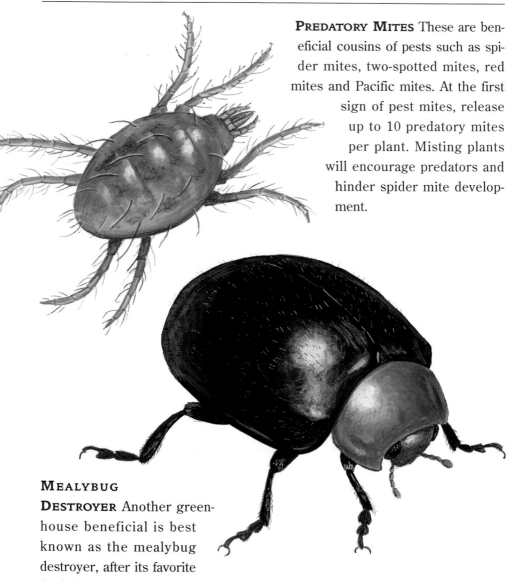

PREDATORY MITES These are beneficial cousins of pests such as spider mites, two-spotted mites, red mites and Pacific mites. At the first sign of pest mites, release up to 10 predatory mites per plant. Misting plants will encourage predators and hinder spider mite development.

MEALYBUG DESTROYER Another greenhouse beneficial is best known as the mealybug destroyer, after its favorite food. It's a non-native lady beetle that won't survive cold winters outdoors. For a single infested indoor plant, release about five beetles. Trap the mealybug destroyers around the infested houseplant by draping sheer fabric over the plant and tying it shut around the pot.

While it may seem odd at first to think of bringing insects into your home and letting them run wild, they'll mostly be polite guests and go about their business of eating what's bugging your plants.

NATURAL PESTICIDES

BY WHITNEY CRANSHAW

Most yard and garden pesticides available today are synthesized chemicals. Because some of these pesticides have been associated with health and environmental hazards, interest in alternatives has been increasing in recent years.

Microbes are one natural source of pest control. Many bacteria, fungi and other organisms cause diseases that kill or cripple insect pests. These are called microbial pesticides.

Combinations of natural products have also proved effective pest controls. For example, combining vegetable oils with an alkaline substance, such as potassium hydroxide, produces soaps that can be used to control mites and insects. These products are known as insecticidal soaps. Vegetable oils or, more commonly, refined petroleum oils, yield horticultural oils that can be highly effective for pest management.

Plants themselves have proven to be sources of some of the most potent pest-control products. Many plants produce a host of chemical defenses that they use to naturally resist attack from various pests. Some plants are especially rich in chemicals that can be extracted and used for insect control. These products are known as botanical insecticides or, simply, botanicals.

Pesticides derived from natural sources, like those that are manufactured from petrochemicals, have a wide range of effects. Most botanical pesticides, for example, do less ecological damage than synthetics because they break down rapidly when exposed to heat, light and water. Others are as acutely toxic (or sometimes more toxic) than common synthetic garden pesticides. (Acute toxicity is a measure of the damage they can do to *you* if they're ingested, inhaled or absorbed through the skin.) All pesticides — synthetic and natural — are regulated as pesticides by the Environmental Protection Agency and the states, and by law must be used strictly in accordance with all instructions on the product labels.

Many bacteria, fungi and other microbes cause diseases that kill or cripple insect pests. The most common microbial pesticide is *Bacillus thuringiensis*, better known as *Bt*.

MICROBIAL PESTICIDES

The microbe most commonly used for garden pest control is the bacterium *Bacillus thuringiensis*, better known as *Bt*. Different strains of *Bt*, which occurs naturally in soils around the world, produce toxins that affect different insects. For example, the "kurstaki" strain of *Bt (Btk)* kills caterpillars, such as gypsy moth, hornworms and cabbageworms. Leaf beetles, such as the Colorado potato beetle, are susceptible to the "tenebrionis" strain *(Btt)*.

To be effective, *Bt* must be eaten by the pest. Susceptible insects stop eating soon after ingesting *Bt*, as it destroys the lining of their gut. Death often follows in a few days.

The primary advantage of *Bt* is its highly selective action. Most *Bt* products only kill caterpillars that eat it. This means that most beneficial insects are spared the adverse effects. *Bt* is considered quite safe to humans, and most products can be used right up to harvest.

Bt does have some limitations, however. Because it must be eaten, thorough coverage of the affected plant is critical. *Bt* also breaks down rapidly upon exposure to sunlight and water, rarely lasting more than a few days. What's more, *Bt* kills not only the larvae of pests, but also the larvae of butterflies. Know the insect you are spraying for and keep away from plants where butterfly larvae feed.

Another bacterium long used for insect control is *Bacillus popilliae*, which produces the "milky spore" disease of Japanese beetle grubs, a major lawn pest. This bacterium has become distributed widely throughout eastern North America, originally through government programs and subsequently by the insects themselves. Milky spore is sold through many garden catalogs.

Microbes may sometimes be used for indirect control of plant pests. Perhaps the best example is the product Clandosan, sold for control of nematodes. This product consists primarily of crab shells and related material that is mixed with the soil. It stimulates the growth of microbes that feed on chitin, the main component of crustacean shells. Nematodes also are covered with chitin and so are susceptible to the microbes as well.

Microbial pesticides are applied as sprays, dusts or granules, just as conventional pesticides are.

HORTICULTURAL OILS

Refined petroleum oils have long been used for managing insects and mites. Oils smother insects by plugging the orifices, called spiracles, through which they breathe. They may also be toxic to some insects and mites. Oil products developed for use in pest management are typically referred to as either horticultural oils or dormant oils. They are usually used as sprays, mixed with water in a 1 to 3 percent solution.

The primary problem with oils is that they can damage your plants if used improperly. The first oils to be used horticulturally were the "dormant oils," which could only be used safely on plants in a dormant state. However, our understanding of what makes oils useful as pesticides and what causes plant injury (phytotoxicity) has increased, and oil products that can be used safely on many plants, even when they've leafed out, are now available. Some plants, such as walnut, certain maples and cedar, do remain "oil shy" even to the most refined

Horticultural oils smother insects and mites by plugging the orifices through which they breathe.

horticultural oils. Read the label for details on sensitive plants and on when *not* to spray (spraying under certain environmental conditions can cause injury).

Oils have remained a popular pest-management option because they are quite effective for many difficult problems. They're most commonly used as dormant sprays to control insects and mites that spend the winter on trees and shrubs. However, the refined oils now on the market are also useful for controlling whiteflies, young scales, mites and many other plant pests present during the growing season. Oils have also proven useful in managing some plant diseases.

Horticultural oils are considered quite safe to humans and other wildlife. Adverse effects on beneficial organisms are also minimal, particularly those of dormant season sprays, which are applied when most beneficial insects are not yet present in the garden. Like soaps, horticultural oils act strictly through contact action and have no residual effects, so thorough coverage is essential.

INSECTICIDAL SOAPS

Soaps have been used as insecticides for over 200 years, but recently their use has increased exponentially. This is largely because there is now a better understanding of which types of soaps make the most effective insecticides, yet do not damage plants.

Insecticidal soaps are applied as dilute sprays (1 to 3 percent concentration) and work primarily by damaging the cell membranes of insects and mites. A wide range of insects are sensitive to soaps — primarily small, soft-bodied species such as aphids, leafhoppers and spider mites. But some larger insects, such as Japanese beetles, are also susceptible. Effects are rapid, usually resulting in death of susceptible insects within a few minutes after exposure. Soaps are sometimes sold in mixtures with other insecticides, such as pyrethrins, to increase their effectiveness.

Insecticidal soaps work against a wide range of insects but must be applied directly on the pests to be effective.

The selective action of soaps and their high degree of safety to humans are their major advantages. Generally, they have a minimal impact on beneficial species. (One significant exception is that soaps kill predator mites, often an important control of spider mites.) Most insecticidal soaps are registered for use on a wide range of vegetable and ornamental plants.

One of the main limitations of soaps is that they work strictly on contact and have no residual effects. This means that they must be applied directly on the target pests, and so good spray coverage is essential. Also, soaps are more sensitive to certain environmental conditions than other insecticides are. For example, the minerals in hard water react with soaps to reduce their activity. And soaps may be less effective if applied during periods when they dry very rapidly.

Although the insecticidal soaps have been developed with plant safety as a major consideration, some plants are sensitive to soaps and can be injured. Most of these are listed on the product labels under the section outlining hazards associated with use. Indeed, research has also identified soaps that are particularly injurious to plants, and these herbicidal soaps are now marketed as contact "weed killers."

Many household soaps and liquid dishwashing detergents can be used effectively as insecticides. These should be applied as dilute sprays. Their main disadvantage is that their effects on plants and insects have not been tested and there is a greater chance that they'll cause accidental injury to your plants.

ALCOHOL

Alcohol affects many types of insects, apparently by causing them to dry out and die. Although there are no commercial alcohol insecticides, alcohol is an ingredient in some insecticidal soaps and "ready-to-use" insecticides.

Alcohol is often used to control mealybugs on houseplants. Usually, it is applied directly onto the insects with a cotton swab in order to avoid injuring the plant. However, many plants tolerate alcohol well, and insects can be controlled with a spray of alcohol and water in equal parts. Try spraying the alcohol solution on a small part of the infested plant first to make sure it does no damage. If after a few days the plant shows no adverse effects, go ahead and spray the entire plant.

Botanical insecticides are poisons derived from plants. Most botanicais do less ecological damage than synthetics do because they break down comparatively rapidly when exposed to heat, light or water. But they are still poisons and must be handled with caution.

BOTANICAL INSECTICIDES

PYRETHRUM The most widely used of the botanical insecticides are extracts from the flowers of the pyrethrum daisy, *Chrysanthemum cinerariifolium*. Powdered pyrethrum flowers are rarely sold for pest control, but there are numerous products containing the extracted active ingredients, pyrethrins. Formulations sold for garden use often combine pyrethrins with other ingredients such as soap, diatomaceous earth or rotenone, another botanical insecticide.

Pyrethrins have some unusual insecticidal properties. Perhaps most striking is the rapid "knockdown" effect they have, which causes most flying insects to drop almost immediately upon exposure. Pyrethrins are also highly irritating to insects and can therefore be used as a "flushing agent" to disperse pests. They also rapidly degrade when exposed to light or moisture and so do not persist for long in the environment.

Most insects are highly susceptible to pyrethrins, so quite low concentrations are applied. At the same time, pyrethrins are quite non-toxic to most mammals, making them among the safest insecticides in use. The short persistence and low toxicity of pyrethrum-derived insecticides have enabled federal regulators to permit their use on a wide variety of crops, typically with little or no interval required between application and harvest. Pyrethrins also are among the few insecticides that are cleared for use around food handling and preparation areas.

In the past few decades, synthetic pyrethrins, or pyrethroids, have been developed. The pyrethroids have the basic chemistry of pyrethrins but are synthetically modified to improve persistence, insecticidal activity and other features. Few pyrethroids are available for yard and garden use, although they are used widely in commercial agriculture.

ROTENONE is one of the oldest botanical insecticides. Records suggest that it was first used against insects in 1848. (For centuries before that it was used as a fish poison.) Most rotenone is derived from South American species of the genus *Lonchocarpus*. Rotenone is used most commonly as a dust prepared by grinding the plant roots or extracting the active ingredients and coating dust particles. Several rotenone/pyrethrins mixtures are marketed.

The Environmental Protection Agency has permitted use of rotenone on a wide variety of vegetables and small fruits. It is both a contact and stomach poison to insects. Rotenone is used primarily for control of various leaf-feeding caterpillars and beetles, such as cabbageworms and Colorado potato beetle. Some insects with sucking mouthparts, such as aphids and thrips, are also susceptible to rotenone. It is a relatively slow-acting insecticide, often requiring several days to actually kill susceptible insects, although they stop feeding shortly after exposure.

Gardeners should be aware of the fact that rotenone is the most acutely toxic of the widely available botanicals — more toxic than most common synthetic pesticides. It is moderately toxic to most mammals, and highly toxic to fish and aquatic life.

RYANIA is the powdered extract from the roots and stems of the shrub *Ryania speciosa*, native to South America. It is sold primarily as a wettable powder. Ryania is also available in some combination formulations with pyrethrins and rotenone.

Ryania has shown promising insecticidal action against many insects. It is sold primarily for control of codling moth. Many caterpillars, leaf beetles and thrips also are susceptible to ryania extracts. Ryania affects these insects either on contact or when eaten. However, it has minimal effects on many beneficial insects, so it can be used with these and other biological controls. Ryania breaks down more slowly after application than other botanical insecticides. It is considered relatively non-toxic to mammals.

SABADILLA is an insecticide produced by grinding the seeds of the sabadilla plant, *Schoenocaulon officinale*. For several years sabadilla products were not available in the U.S. However, they are now sold by several mail-order suppliers, and garden centers have begun to carry sabadilla products as well.

Sabadilla is both a contact and stomach poison and has shown greatest promise against several of the "true bugs," such as squash bug, chinch bug, harlequin bug and stink bugs. It has proven effectiveness against leaf-feeding caterpillars, Mexican bean beetles and thrips. Use of sabadilla on certain vegetables, including squash, cucumbers, melons, beans, turnips, mustard, collards, cabbage, peanuts and potatoes, is permitted by the EPA..

The ground seeds of sabadilla sold for garden use are considered among the least toxic of the various botanicals. However, sabadilla dusts can be highly irritating to the respiratory tract, often provoking a violent sneezing reaction if inhaled. Be sure to wear a dust mask when applying it and, as with all pesticides, follow precautions listed on the product labels. In addition, several of the alkaloids in sabadilla can cause rapid depression of blood pressure in mammals.

NEEM The newest of the botanical insecticides are those derived from seeds of the neem tree, *Azadirachta indica*. Extracts from neem seeds and other parts of the tree have long been used for pharmaceutical purposes, for example in toothpaste, particularly in India. Recently, neem has received a great deal of attention because it is so safe to humans and has unusual properties against insects.

Sprays of neem applied to leaves often deter feeding. Furthermore, neem apparently affects the hormones many insects need to develop, killing them as they attempt to molt or emerge from eggs. Many leaf-chewing beetles and caterpillars can be controlled with neem insecticides. Aphids and most other sucking insects generally are less susceptible.

Because of its demonstrated safety, neem was recently exempted by the EPA from food-crop restrictions, enabling manufacturers to market it for use on any edible or ornamental plant.

HOW TO READ A PESTICIDE LABEL

Product labels are the home gardener's most important source of information about a pesticide's effect on the environment and human health. A label gives you the most up-to-date information on everything from how to mix and apply a pesticide, to which active ingredients it contains, to which pests it can kill. It is against federal law to use any pesticide in any fashion not in accordance with instructions on the label — a fact that is often overlooked by home gardeners. The annotated label below points out what information is available on a pesticide label, and where.

BACK PANEL

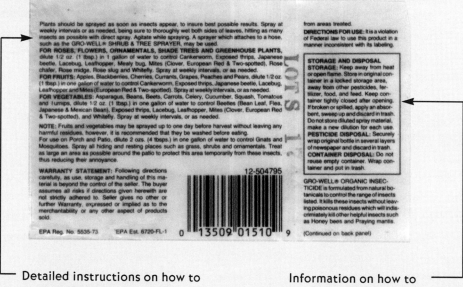

Detailed instructions on how to mix and apply the pesticide and a list of all pests and crops for which the Environmental Protection Agency permits its use.

Information on how to store the pesticide and dispose of the container.

The chemical that kills the pests, listed as a percentage of the formulation.

Cautions regarding the use of the pesticide and its human health hazards.

FRONT PANEL

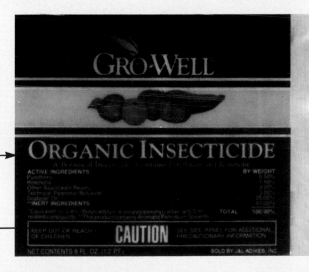

Signal word, listed in large letters, indicates toxicity:

CAUTION
relatively non-toxic or slightly toxic

WARNING
moderately toxic

DANGER/POISON
highly toxic

A list of all known hazards to wildlife, beneficial insects, groundwater, etc., plus general instructions on avoiding these hazards.

S U P P L I E R S

The following companies offer a variety of natural pesticides, traps, beneficial insects and other products:

BRICKER'S ORGANIC FARMS, INC.
824 Sandbar Ferry Rd.
Augusta, GA 30901
(800) 200-5110

EARLEE, INC.
2002 Highway 62
Jeffersonville, IN 47130
(812) 282-9134

GARDENS ALIVE!
5100 Schenley Pl.
Lawrenceburg, IN 47025
(812) 537-8650

GARDEN-VILLE
8648 Old Bee Cave Rd.
Austin, TX 78735
(512) 288-6115

HARMONY FARM SUPPLY
P.O. Box 460
Graton, CA 95444
(707) 823-9125

INTEGRATED FERTILITY MANAGEMENT
333 Ohme Gardens Rd.
Wenatchee, WA 98801
(800) 332-3179

MELLINGER'S, INC.
2310 West South Range Rd.
North Lima, OH 44452
(216) 549-9861

NATURAL GARDENING COMPANY
217 San Anselmo Ave.
San Anselmo, CA 94960
(415) 456-5060

NATURE'S CONTROL
P.O. Box 35
Medford, OR 97501
(503) 899-8318

NECESSARY TRADING COMPANY
422 Salem Ave.
New Castle, VA 24127
(703) 864-5103

PEACEFUL VALLEY FARM SUPPLY
P.O. Box 2209
Grass Valley, CA 95945
(916) 272-4769

PEST MANAGEMENT SUPPLY INC.
311 River Dr.

Hadley, MA 01035
(413) 549-7246

RINGER CORPORATION
9959 Valley View Rd.
Eden Prairie, MN 55344
(612) 941-4180

SAFER INC.
9959 Valley View Rd.
Eden Prairie, MN 55344
(800) 423-7544

Photo Credits

Cover and pages 1; 8 upper left, center right and lower left; 77; 80; 82; 84 top and center; 88; 89 by judywhite

Pages 8 upper right and center left; 84 bottom by Pamela Harper

Page 11 by Pam Pierce

Page 91 by Elvin McDonald

Page 8 lower right by James Zablotny

Pages 96; 98; 99; 101 by Ariel Jones

INDEX

American Cottage Gardening

Annuals: A Gardener's Guide

Bonsai: Special Techniques

Culinary Herbs

Dyes from Nature

The Environmental Gardener

Ferns

Garden Photography

The Gardener's World of Bulbs

Gardening for Fragrance

Gardening in the Shade

Gardening with Wildflowers & Native
    Plants

Greenhouses & Garden Rooms

Herbs & Cooking

Herbs & Their Ornamental Uses

Hollies: A Gardener's Guide

Indoor Bonsai

Japanese Gardens

The Natural Lawn & Alternatives

A New Look at Vegetables

A New Look at Houseplants

Orchids for the Home & Greenhouse

Ornamental Grasses

Perennials: A Gardener's Guide

Pruning Techniques

Roses

Soils

The Town & City Gardener

Trees: A Gardener's Guide

Water Gardening

The Winter Garden

21st-Century Gardening Series

For centuries, gardens have been islands of beauty and tranquility in an often disorderly, unpredictable world. The late-20th-century garden is also a major arena in the struggle to balance human and ecological needs, one of the great tasks of our time. Brooklyn Botanic Garden's 21st-Century Gardening Series explores the frontiers of ecological gardening. Each volume offers practical, step-by-step tips on creating environmentally sensitive and beautiful gardens for the 1990s and the new century.

TO SUBSCRIBE OR ORDER:

21st-Century Gardening Guides are published quarterly — spring, summer, fall and winter. A four-volume subscription is included in BBG subscribing membership dues of $25 a year. Mail your check to Brooklyn Botanic Garden, 1000 Washington Avenue, Brooklyn, NY 11225.

For information on how to order any of the handbooks listed at left, call (718) 622-4433.